Transdisciplinary Research Across Knowledge Worlds

Stanley D. Brunn
Editor

Transdisciplinary Research Across Knowledge Worlds

Beyond Parallel Universes

Springer

Editor
Stanley D. Brunn
Department of Geography
University of Kentucky
Lexington, KY, USA

ISBN 978-3-032-19490-9 ISBN 978-3-032-19491-6 (eBook)
https://doi.org/10.1007/978-3-032-19491-6

This Springer imprint is published by the registered company Springer Nature Switzerland AG
The registered company address is: Gewerbestrasse 11, 6330 Cham, Switzerland

Preface

In contemplating, constructing, and editing a book on a new interdisciplinary and international topic, there are always individuals who merit acknowledgment. They will come from previous generations who influenced early thinking about the topic and also present generations of friends who provided thoughtful guidance about possible ideas and contributors. The "parallel universe" concept while not new is a subject that might be initially applied solely to astronomers. However, on further reflection it can be applied to "earth worlds" in many contexts. There is no doubt that there are many parallels on earth, not only considered in cultural, economic, social, and environmental contexts, but at many political scales. These parallel worlds can and do exist in people living side by side, but also those separated by long distances. Those distances may be class, race, religion, and technology but also age, gender, heritage, knowledge acquisition, and practice. Parallelism can be observed in a "place or space" context, but also applied to "time" which, like space, has both degrees of separateness and also parallelism.

The foundations for early thinking about parallel universes came from individuals who have had impacts on how I looked at various worlds in the past and how I envision/consider them in the present. Some were former professors who taught classes I took and liked, others were scholars whose contemporary writing impacted my thinking, and still others were from previous decades and generations. Among the early geographers whose writings have contributed to my lifetime learning and re-learning about human and environmental worlds are these individuals (in alphabetical order): Bill Bunge, Anne Buttimer, Karl Butzer, Peter Gould, Peter Haggett, David Harvey, John House, Donald Meinig, Janet Monk, Richard Morrill, Fred Simoons, O.H.K. Spate, Ned Taaffe, Peter Taylor, Waldo Tobler, Yi-Fu Tuan, Bill Warntz, Gilbert White, Don Zeigler, and Wilbur Zelinsky. All were scholars who encouraged "outside the box thinking and writing" that expanded our "comfort knowledge zones" (yes, plural).

Added to the list of inspiring professionals are those I contacted several years ago when the parallel universe concept was surfacing. Many not only suggested topics, but named potential authors and reviewed submitted chapters. They shared the need for geographers to explore new knowledge universes that would both broaden and deepen our thinking about topics, theories, and models that were more interdisciplinary, cross-disciplinary, and international. At the same time, these new worlds would present challenges for those thinking about ideas for new maps and mapping. These friends include, in alphabetical order, Karen Barton, Tom Bell, Martin Dodge, Ronnie Donaldson, Ghazi Falah, Don Janelle, Jim Johnson, Peter Jordan, David Knight, Lily Kong, Virginie Mamadouh, Kent Mathewson, Alec Murphy, David Newman, Karl Raitz, Dennis Rumley, Jorn Seeman, Herman van der Wusten, and Barney Warf.

I also want to thank Matt Wilson, chair of the University of Kentucky Department of Geography, for providing the office space for an emeritus professor where I could work on many on-campus tasks that led to this book seeing the light of day. He has always been a colleague who recognized the importance of seeing new ideas and new practices expressed by youthful or senior faculty.

As the book idea was taking shape, I was thinking about the best publisher to share the idea and the prospectus with. That was not a difficult decision as I shared the proposal with Evelien Bakker at Springer. She liked the idea and proposed a better title. It was yet another book with Springer. For more than two decades Evelien has always been a supporter of innovative book topics that included working with scholars from various disciplines and countries across the world. Prasad Gurunadham was a second professional at Springer who has also been good to work with on multiple publication tasks in the past decade.

Preparing all the chapters for the publisher is a very important part of any publication that reaches the market for readers around the world. The person I am most indebted to for these tasks is Donna Gilbreath who has done the same for other Springer books published in the past two decades. Her professional manuscript preparation and editing skills are outstanding. She also enjoys learning new geographies and working with authors around the world.

Finally, I want to thank my wife, Natasha, for her support on this idea from start to finish. She always listened to new ideas and novel approaches to expanded geography worlds. Also, she recognized the importance I placed on interdisciplinary topics and working with international authors around the world, a task that brought daily new learning about new parallel worlds and thinking. She learned much about the concept, the editing process as well as the editor.

Lexington, KY, USA
October 2025

Stanley D. Brunn

Acknowledgments

We want to thank Donna Gilbreath for working with authors to ensure their chapters are of highest quality publication. Her patience, steadfastness, and experience are among many reasons why the volume is a contribution to multiple disciplines and knowledge frameworks.

Contents

Contributors

Krasimir Asenov University of Agribusiness and Rural Development (UARD), Plovdiv, Bulgaria

Karen S. Barton Geography, GIS, and Sustainability, University of Northern Colorado, Greeley, CO, USA

Thomas L. Bell Department of Geography and Sustainability, University of Tennessee, Knoxville, Tennessee, USA

Lori Boyer University of North Carolina, Chapel Hill, NC, USA

Stanley D. Brunn Department of Geography, University of Kentucky, Lexington, KY, USA

John A. Cross Department of Geography, University of Wisconsin, Oshkosh, WI, USA

Cassandra R. Davis Department of Public Policy, The University of North Carolina, Chapel Hill, NC, USA

John DePriest Center for Global Education, Tulane University, New Orleans, LA, USA

Simona Goldin Department of Public Policy, The University of North Carolina, Chapel Hill, NC, USA

Omari Graham The University of the West Indies Seismic Research Centre, St. Augustine, Trinidad and Tobago

R. D. K. Herman The Pacific Worlds Institute, Hilo, HI, USA

Nygmet Ibadildin Department of Political Science and International Relations, College of Social Sciences, KIMEP University, Almaty, Kazakhstan

Iheoma Iruka Gillings School of Global Public Health, The University of North Carolina, Chapel Hill, NC, USA

Stephanie Mahin University of North Carolina, Chapel Hill, NC, USA

Irena Nikolova University of Agribusiness and Rural Development (UARD), Plovdiv, Bulgaria

John Pickles Department of Geography and Environment, University of North Carolina, Chapel Hill, NC, USA

Iain Ross Cajamarca, Peru

Dundas, ON, Canada

Allison Schlobohm University of North Carolina, Chapel Hill, NC, USA

HaeRan Shin Department of Geography, Institute for Korean Regional Studies, Institute for Gender Studies, Institute for Peace and Unification Studies, Seoul National University, Seoul, Republic of Korea

Gabriellę Thongs Department of Geography, Faculty of Food and Agriculture, The University of the West Indies, St. Augustine, Trinidad and Tobago

Arlette Saint Ville Department of Geography, Faculty of Food and Agriculture, The University of the West Indies, St. Augustine, Trinidad and Tobago

Introduction

Stanley D. Brunn

Planet Earth has many worlds. These include worlds of culture, economic livelihood, ethnic and racial diversity, social well-being and environmental settings, but also political conflict, media reporting, migration, health care and healing. These and many other worlds exist side-by-side as well as short and long distances away. These differences can be explored, interpreted and even mapped in familiar/traditional ways or in novel ways using electronic media. These worlds for individuals, groups and countries may be visible and invisible in daily life as well as in contemplating and planning for present and future personal and global settings.

Much of our understanding of these diverse worlds of similarities and contrasts comes from what we observe. That is, people-gatherings of those who work, reside, worship and entertain are often side-by-side those who have similarities in working, worshipping, social class origins, race and ethnic features and political thinking.

To the observer, one may think that the "look alikeness" is commonplace. Those likenesses may appear on maps showing a common feature about a city's or countryside's culture, economy, politics and heritage. Closer inspection of who is living where or who is networking with whom or in close contact with others may reveal some rich diversities of those living close to them. That is, while outwardly there may appear to be much sameness, a closer read will reveal there are many differences—differences in ways of living, working, worshipping, entertaining, social media using and even in heritage backgrounds, gender identities and roles, political thinking and religious/secular practices and preferences.

The previous paragraph describes a world of possibly much diversity that is not apparent in an observable or visible context. However, a casual or shallow interpretation may provide some slight differences but a deep inquiry into who lives, works, and worships where or with whom one interacts personally or via social media or

S. D. Brunn (✉)
Department of Geography, University of Kentucky, Lexington, KY, USA
e-mail: stan.brunn@uky.edu

S. D. Brunn (ed.), *Transdisciplinary Research Across Knowledge Worlds*,
https://doi.org/10.1007/978-3-032-19491-6_1

the internet may reveal significant differences. These shallow or deep worlds may reveal little-known or unknown differences not initially observable but are evident on close examination. It also may be that there are "known" worlds of differences but that they were not studied previously by scholars because they did not think they were important or, perhaps, they did not know how to study them. Regardless of the reason for neglecting them, two major points are important. One is that these different shallow and deep worlds exist and, second, that they merit examination in different contexts.

Parallel Worlds Exist

These parallel or non-intersecting worlds are important features of understanding the cultures, economies, social well-being, and politics of both people and places on the planet. They not only exist, some with lengthy histories, others in current or recent times, but they also persist. Cultures and populations are not all alike in their heritage, assimilations, social features or political thinking. This thinking has foregrounded much of the research by social and behavioral scientists as well as those in the humanities who have studied specific features of historical and contemporary groups in disciplinary contexts. Historians, geographers, sociologists, economists, anthropologists, psychologists and others have devoted much time to studying specific groups, cultures, and economies about their origins, migrations and ways of living, working, thinking and assembling.

What has been lacking in these lengthy research efforts is detailed investigations into how these different groups interact or get along together and live together in spite of differences. Those "getting along" efforts occur with those living close to or side-by-side others who may have different degrees of working, thinking, worshipping, entertaining and networking. The social, cultural, and economic interactions exist in varying degrees, from little or slight to significant or almost bipolar.

Understanding these "getting along" efforts among those in different cultures and heritages comes not only from learning about groups of people working or living alongside others, but also their celebrating, worshipping, entertaining and networking, but also where (places) and how their daily life and living intersect with others. All human activity occurs in places and spaces, close and distant, as well as time settings, deep or shallow pasts or contemporary. Those different time/place experiences and worlds may be little known or unknown, but they exist in the 24/7 lives of everyone. Recognizing this point brings the construction and interpreting of maps and mapping as central to providing a better understanding of these different or alternate worlds within a community, whether it be a rural, suburban or urban setting. Maps show where something is occurring and who is in those spaces and places. Those places may be people much alike, somewhat alike or much different. Those places may be the same every day or nearly the same or they may be different in time—morning or afternoon, weekday or weekend, monthly or seasonally. They may be different for youth and elders, for women and men, for new and lifelong residents, for those in poverty and gated communities, and for those in one political party versus another. There may be clusters or "much sameness" among residents or clusters of much or similar diversity. The e-worlds of communication and interaction have ushered in an additional dimension of people's definition of community. It is one where closeness is defined not by proximity, but by networking. That networking community may be with others living nearby physically or in another city or suburban location or living and working in another time zone, country or continent.

The Emergence of Parallelism

The previous paragraph describes the concept of "parallel universes" and how it can and might be applied to looking at cultures, economies, social well-being, ethnic groups, religions and politics in cities as well as rural areas. Parallel in a strict context is a concept that can be applied to worlds where there is little or no interaction between one world and another. Those "worlds" may be where groups (economic, social, political, etc.) exist side-by-side, physically as well as cognitively, and do not, for various reasons, intersect or interact with each other. Examples of the parallelism concept can be applied to the following;

- Low- and high-income residents live in separate places and universes.
- Religious believers of different heritages and backgrounds do not interact.
- Women and men do not mix together socially.
- Lifelong and new refugee residents live in alternate worlds.
- Social media networking is invisible and ignores space and place.
- Race and ethnic members are networked and separated not living in clusters.
- Lifelong residents and seasonal tourists interact in separate place universes.
- Language communities in metro areas are like isolated archipelagos.
- Indigenous and colonial heritage newcomers are separated.
- E-worlds include individuals networking both "on" and "off" the grid.
- Religious communities are "separate ghettos" from other faiths.
- Places and spaces where "parallelism" emerges, exists and persists.
- Places and spaces where "parallelism" is declining: why and how.
- The role of the state, NGOs and mass media in parallel universe existence.

Mapping the Universes

Keys to understanding the social and geographical degrees of parallelism that persist come from scholars examining five distinguishing features of the "separated" worlds and groups, viz., why they exist and persist. These can be applied to groups in both urban and rural settings. First, what are the distinguishing features of those two or more worlds (ethnic, racial, religious, social, etc.) that are parallel? Second, are there specific differences in the "parallelism" between groups. For example, are there deep cultural or technological differences? Are there distinct physical features, such as walls, entrance points, language signage, and physical barriers? Third, what steps are taken by the state or a specific group to maintain the separateness or to reduce the parallelism that exists? Fourth, what maps can/might be prepared that show the differences and degrees of differences between and within specific groups? These might be maps showing visible physical features or cognitive differences between children and adults, women and men, old timers and new arrivals. Fifth, and in an increasingly visual-oriented world, what photos can be taken to show the parallelisms that exist. These may be aerial photographs, views along a street, facing houses and shops across a street, or entrance signs to the parallel settings. Visibility is an important part of parallel universe features in urban, suburban and rural areas.

Bringing the Concept to Real Life

The 14 chapters both deepen and broaden our understanding of parallel universes by exploring some innovative approaches and topics in different human and human/environmental settings. Pickles applies the concept to geography's disciplinary thinking from the 1950s to the 1980s, a time when geography was moving from descriptive thinking to a quantitative theoretical world followed by humanistic geography's emergence. The next several chapters apply the concept to cultures that lived side-by-side, but had some sharply different cultural heritages. Cross applies the concept to Old Order Amish and modern farmers in rural Wisconsin. Shin discusses the concept to those from North Korean moving into South Korea. Asenov and Nikolova apply it to separate the Roma communities in Bulgaria, and Ibadildin to the Kazakh and Russian communities in Kazakhstan. Religious communities provide additional examples of "separateness" in their practices. Ross discusses these communities in contemporary Cajamarca, Peru and De Priest applies these worlds from English hymns translated into Choctaw. Two focus on racial issues in the United States, one by Schlobohm et al. on historical segregation in the South and other by Davis et al. on Freedmen's towns which are subject to natural disasters. Climate change issues are brought to light by Barton in her discussion about how a green energy paradox exists between wind farm developers and Sami voices in Norway. The indigenous voices are front and center in Herman's presentation about indigenous groups in the Pacific using the World Wide Web as an important visual source of identity and meaning. Bell's discussion of the visual worlds is explored through U.S. postage stamps which have become a much more visual source about heritage, a shift he labels as one from patriotism to pandering. In the final chapter, Graham et al. illustrate how important visual sources, viz., photos and maps, have played key roles in Montserrat's coping with volcanic emergencies.

Where We Go from Here

The chapters provide a rich source for ongoing research in various fields of study, not only human and environmental geography, but in the social and natural sciences as well as in the humanities. All provide examples of how important concepts about place, space, environments, institutions and politics are in understanding both the visual and invisible worlds surrounding humankind in various human/environmental settings. They also demonstrate the importance of capturing some of those interfaces and interrelationships in photos and especially maps. Visualization is a key and overriding concept. There is no question but that parallel universes merit further study by scholars in separate social, behavioral and environmental sciences, those venturing into intersecting conceptual and theoretical territories, but also those in the humanities, whether focusing on art, music, drama, humor, storytelling or mass media. Breakthroughs are sought both within and between the knowledge worlds at personal, local, regional and global scales. Following are some suggested topics for further research by scholars working within a specific field or discipline or willing to engage in some venturesome cross- and interdisciplinary topic or conceptual framework:

- the networked worlds of new residents moving into new home locations,
- the political, social and technological boundaries of those "on and off" the grid,
- the first weeks of moving "from off to on" the grid
- the daily personal/family life in wealthy gated communities and adjacent slums,
- solo and multiple language worlds of children, adults, and refugees,
- the social networking (local to global) of rural residents in small towns,
- the increased or decreased importance of maps on cyberworlds,
- the role of the state in increasing or decreasing social parallelism,
- the archipelagos (rural and urban) of connectedness and disconnectedness,
- the cyber-binary worlds of temporary cruise ship travelers and on-shore residents,
- the parallel worlds of those with hearing and seeing losses,
- the daily time/place journeys of undocumented residents, exploring "grayness,"
- the boundary worlds of very high to no tech communities and individuals,
- comparing the "voices" of "the heard" and those "in silence,"
- from no sight to gaining sight; and from having to losing sight;
- gender differences in leadership of caring and non-profit organizations.

Challenges Awaiting

A careful or casual reading of the following chapters should stimulate the reader to think about the concept of parallel universes and how it can be applied to studying a wide variety of topics in the social and behavioral sciences and humanities. Research on many traditional and familiar topics in various disciplines is now being expanded beyond traditional disciplinary borders and boundaries. What is behind much of this re-thinking is addressing topics that are inherently inter- and transdisciplinary in content. Examples that readily come to mind are climate change, COVID-19, gender and racial issues, sustainability and mass media. Someone studying these will find that focusing solely on a single discipline's methods, models, terms, literature and conceptual frameworks is too narrowly focused. Instead, scholars are finding that there needs to be much more appreciation, acceptance and recognition of what other fields or disciplines have to offer. This point, and one evident in the book's chapters, is an additional example of where the knowledge worlds are moving from some "parallel" thinking and acceptance to "intersecting" thinking. These emerge and explore new, and often unheard of concepts, terms and theories. We can expect to see these intersecting worlds emerging at various rates and ways across the knowledge worlds including the sciences and humanities, work and leisure, the spiritual and secular, the political and non-political, gender and race, and text and visual. Already, these new worlds are evident in the pioneering inter- and transdisciplinary training of professionals, the labels of majors and minors in universities, names and foci of new professional journals, research funding priorities, and employment descriptions. In truth, the moves from parallel to intersecting worlds are much in evidence.

Geography's Intersecting Paths of Science and Humanism: 1950s–1980s

John Pickles

Introduction

What a time it was! A trans-Atlantic regime under the financial and military protection of the US led to the 'thirty glorious years' ('*les Trente Glorieuses*'); a post-war boom in which a state-capital-labor compact created a virtuous cycle that expanded state administration and manufacturing, allowed for expanded consumption of goods, including new ways of experiencing the world through new urban forms, media and tourism (Lefebvre, 2022). Social democratic ideals permeated trans-Atlantic society as war survivors and post-war youth responded to the horrors of the war and the camps it created. It was also an atomic era in full bloom, science and technology promised uncertain futures while Cold War geopolitics scripted every aspect of international life and national budgets. Domestically urban 'development' remade city lives under the umbrellas of improvement, modernization, and efficiency. At the same time, the dismantling of old imperial and colonial orders opened new arenas of experimentation with revolution, nationalism, state building, and social and economic development.

Not surprisingly geography responded, albeit unevenly. Wartime survivors and GI bill-students in the US entered universities armed with new statistical tools, algebraic models, and a sense that science and rational planning could change the world (Ackerman, 1954). In the US and UK, these were the 'space cadets' who challenged what they perceived to be the descriptive banalities of traditional regionalisms. They saw areal differentiation as being too static, too descriptive, and lacking in tools to analyze general patterns and laws of spatial organization. The 'space cadets' were having none of it. They knew the practical value of probabilistic models, the power

J. Pickles (✉)
Department of Geography and Environment, University of North Carolina, Chapel Hill, NC, USA
e-mail: j.pickles@unc.edu

S. D. Brunn (ed.), *Transdisciplinary Research Across Knowledge Worlds*,
https://doi.org/10.1007/978-3-032-19491-6_2

of statistics and geometrical analysis, and the intellectual draw to explore regularities and laws (Gould & Pitts, 2002).

At the same time, the rise of technocratic politics and planning logic in transforming post-war landscapes of city and suburb alike stimulated existential concerns for the conditions of life that these models and practices were creating. Zamyatin's *We* (1924) in Stalin's USSR, Huxley's *Brave New World* (1932) and Orwell's *1984* (1949) warned against the objectifying tendencies of overly bureaucratic and technological regimes, social scientists like Michael Oakeshott (1962) and Jane Jacobs (1961, 1969) questioned the increasing role of rationalism in politics and urban planning, and from 1959 on architects such as Kisho Kurokawa (1988) and his colleagues in the Metabolist movement reacted to post-war rapid economic growth and technological development (particularly in Japan) by arguing that design should differentiate space based on its meaning and lived experience to create forms of 'symbiosis' in their buildings. When Malvina Reynolds wrote and later Pete Seeger released the song 'Little Boxes' (1963) they were expressing a generational sense of ennui in a world increasingly subject to technocratic management and the bland reproducible modernist architecture that resulted (Leatham, 2023). What would a sense of place mean under the emerging regimes of capitalist and bureaucratic abstraction? Who was making decisions about and for the rich diversities of social being that seemed to be emerging from the 1960s? What was happening to the everyday lives of people in the increasingly administered communities in which they lived?

These two parallel trajectories met in geography departments across the US and the UK, sometimes with suspicion of the motives of the other. In this chapter, I outline the swaying back and forth between scientific geographies forged out of strong wartime and post-war commitments to building new democratic spaces and societies and critiques that evoked the necessity of holding on to values, ethics, and morality of what it means to be human and to live a full life, to be a social being in what appeared increasingly to be a technocratic, abstract society. In teasing out these threads of parallel engagement and disagreement, I understand that both the fabric of the story I am telling and the threads that make up the story are themselves highly complex and multiple. To that end, this chapter only gestures to the complexity of relations within and between them, to selected actors in this history, and to parallels in social thought and practice outside of the discipline which shaped ideas and self-understanding within it (see Ferretti, 2020). The discussion then turns to a wider conception of science and humanism that came to dominate the intellectual life of the discipline in the 1980s before concluding with some reflections on the demographic causes of intellectual churn.

The New Scientific Geography

Geographers have long held that the emergence of scientific geography could be read as a reaction to the descriptive geographies of the early twentieth century and a reaction to environmental determinist claims to build a predictive science. In the 'Quantitative Revolution' that emerged in the 1950s and early 1960s geographers were differentially prepared for the power of spatial analytical models they created. Trevor Barnes (2025) has located this project in the UK where:

> From the late 1950s, the physical geographer, Richard Chorley, and the human geographer, Peter Haggett, at the Department of Geography, Cambridge University, fundamentally reconceived the geographical project.... Their 'new geography' would instead be explanatory, theoretical, mathematical, and model-based, applying equally to both human and physical geography. As a result, the traditional two halves of the discipline that were often uneasily connected would be seamlessly joined by a single common scientific method.

That method came to be grounded in a common epistemology through David Harvey's (1969) *Explanation in Geography* which outlined a positivist methodology for the study of space and environment.

This unified scientific method sought to ground the various subject areas in the discipline in a common set of assumptions that spatial patterns were observable, measurable, and could be organized in such ways as to produce general laws (Harvey, 1969). Here, positive science and objectivity were to be achieved by a quantitative analysis of relationships among discrete variables, each enabling modeling and prediction, an approach Arthur Strahler (1952) used to outline a dynamic and unified systems approach for geomorphology. Geographers followed one of several paths into spatial analysis. One path was through statistics and particularly the work of J.Q. Stewart (1941, 1947, 1948, 1950) and his influence on William Warntz (Stewart & Warntz, 1958). Warntz and the Harvard Graphics Lab were particularly influential in their efforts to think of geography as a form of social physics (Barnes & Wilson, 2014) (but also later shaping analytical Marxist geography through Warntz's student Eric Sheppard).

For other geographers, US regional science, particularly with the work of Walter Isard (e.g., 1956) and his colleagues at the University of Pennsylvania, opened new horizons linking spatial science to public policy debates (Warntz, 1957). Regional science drew upon the recently translated works of European economists such as August Lösch (1940) and Walter Christaller (1933, 1966), developing their insights for a more objective analysis of economic location theory (Isserman, 2004), drawing the attention of geographers to regional science, notably Gunnar Olsson who was in residence at U Penn in 1963 and 1964 before taking up his assistant and associate professorships at the University of Michigan 1966 through 1971 (Olsson, 1965, 1968, 1975).

Another important path was through cybernetics in which scholars turned their attention to ideas of systems in which certain kinds of relations and feedback were important in shaping concrete outcomes. Norbert Wiener's (1948, 1950) work on information theory and his anti-aircraft work led him to formulate cybernetics (see

Galison, 1994). Excited by the inter-disciplinary possibilities of cybernetics and systems analysis, the study of "circular causal and feedback mechanisms in biological and social systems" stimulated research across the social sciences (von Foerster et al., 1951). Kenneth Boulding (1956, 1962a, 1962b) was a particularly important bridging figure for geographers through his many books linking general systems theory to evolutionary models of environmental change and to social ethics. The result was a general sense about the possibilities of cybernetics that Margaret Mead (1968, p. 3) interpreted as a form of cross-disciplinary thought which allowed members of different disciplines and countries to communicate with each other through "a language in which … systems could be described in a way that was ideologically neutral." These were ideas that also influenced the attempt to create an interdisciplinary applied science of human settlements–'ekistics'–and its journal *Ekistics* (published 1957–2006).

Like all major shifts in institutions and ideas, the space cadets created new networks of exchange and influence, new departments and programs, and new heroes who articulated exciting possible futures. While the standard texts of university courses in the 1950s, 1960s, and 1970s were variously Richard Hartshorne's (1939) *Nature of Geography* in the US and Robert Dickinson's (1976) *The Regional Concept: The Anglo-American Leaders* in the UK, these insurgent 'space cadets' were building specific hotspots and networks of quantitative geography. Fred Schaefer's (1953) criticism of exceptionalism was an important stimulus to quantitative geography, particularly Bill Bunge's *Theoretical Geography* (1962, 1979; see Barnes & Bergmann, 2022). Published in the Lund Studies in Geography Series, Bunge's work was part of a broader center of important work focused on Lund University and particularly through the influence of Torsten Hägerstrand (e.g., 1953).

There were distinct geographies of these shifts in disciplinary ideas and practice. In the UK, scientific geography revolved around Cambridge and Bristol, while Oxford, for example, remained more traditional in its teaching and hiring. In the US, scientific geographies with the goals of universalism, deduction, and prediction developed most prominently in a network of individuals and universities, while traditions of areal differentiation with their commitments to deep local studies and a focus on the specificity of place and region held sway in other departments. As Browning (1982) and Gould and Pitts (2002) show in their conversations with geographers, spatial analysis was a close-knit network of relatively few individuals, working paper series, workshops, and departments exchanging faculty visitors. The University of Washington Department of Geography was one of the key early centers of spatial analysis, particularly after the arrival of Edward Ullman, William Garrison, Brian Berry, and Richard Morrill in the 1950s, when it became one of the first departments in the US to include quantitative methods and statistical models in its research and teaching, playing a key role in the training of future professional geographers (e.g., Garrison & Berry, 1958). From the universities of Penn State, Bristol, Cambridge, Iowa, Northwestern, Ohio State, and Wisconsin (among others) academics created vibrant networks sharing new models, new forms of analysis, and new approaches to space and society. Word traveled quickly about the possibilities of rational analysis of space. More accurate modelling and prediction became central elements of a science and application to real world contemporary problems became crucial, but so also was the expansion of subject matter and methods deployed, perhaps well illustrated by Peter Gould's extension of his work on spatial diffusion to incorporate the informational spaces of mental maps (Gould, 1969, 1975; Gould & White, 1974). Elsewhere the emerging geography of quantitative geographies was uneven. What Brunn (2004) referred to as the 'midwesternization' of Southern geography departments in the 1960s and 1970s was, in essence, a delayed and uneven adoption of the new scientific approaches. Across the region, Southern departments were among the last to bring in "quantifiers" or more than one (Brunn, 2004). Elsewhere, semi-independent research groups emerged, such as the influential Harvard University's Laboratory for Computer Graphics and Spatial Analysis, founded in 1965 by Howard Fisher and later including William Warntz and Jack Dangermond.

Key texts were important in consolidating and normalizing spatial thinking. In the UK, Peter Haggett's *Locational Analysis in Human Geography* (1965) and Richard Chorley and Peter Haggett's (1967) *Models in Geography* were crucial elements in framing the emerging new geography. In the US, Berry and Marble's (1968) *Spatial Analysis: A Reader in Statistical Geography* and Abler et al.'s (1971) *Spatial Organization: The Geographer's View of the World* were particularly important in synthesizing and illustrating the potential of spatial analysis and providing a resource for a new generation of students. Michael Ray's, 1972 review located Abler et al. (1971, p. 447) clearly in this way:

> *Spatial Organization* makes an important contribution to geography as liberal education by presenting, with profuse examples, some of the ways in which geographers think about spatial structures and processes in human behavior. The approach is nomothetic, and the emphasis is on understanding common elements in the perception and organization of space and not on describing unique attributes that distinguish world regions. The purpose is to present strategies that geographers apply to solve problems created by population distributions and human activity.

Spatial Organization was particularly influential in graduate training. Indeed, it is difficult to think about the emergence of spatial science in Geography without some understanding of the roles played by works such as these in conjunction with the nodes and networks of departments, workshops, and journals through which people and ideas circulated.

For example, *Geographical Analysis* was founded in 1969 at Ohio State University to focus on geographical theory, model building, and quantitative methods. Both quantitative methods and spatial analysis were also underwritten by the growing role played by research funding, particularly from the National Science Foundation where its Geography Division supported research in scientific geography, and opened funding to ever wider methodological approaches, particularly under the administration of Ron Abler, Tom Baerwald, and their colleagues. NSF-funded summer institutes in spatial thinking were also a central element in forging collaborative communities and projects. In the UK, the Study Group in Quantitative Methods of the Institute of British Geographers was a similarly important venue for quantitative geography, especially with the creation of the CATMOG Series

(Concepts and Techniques in Modern Geography) published 1975–1996 and now made available by the Quantitative Methods Research Group (QMRG) at the Royal Geographical Society (with the Institute for British Geographers) at https://qmrg.github.io/CATMOG/.

Humanism

The foundations of humanistic geography were wide-ranging, and its relationship with parallel developments in the wider humanities were complex (Pickles, 1986). Perhaps the most recognizably 'humanistic geographical' work was, and still is, identified with a discrete number of key scholars. Yi-Fu Tuan (1974, 1977) asked geographers to focus on lived experience and *topophilia* (the love of place) in understanding the cultural determinants and variability of meanings that have been ascribed to space and place. Sister Annette Buttimer (1974) challenged geographers to incorporate questions of values and ethics into their studies. Edward Relph's *Place and Placelessness* (1976) focused on the creeping homogeneity of modern development as an existential threat to identity and belonging in places that are becoming ever more 'placeless'. And Ley and Samuels (1978) were particularly influential in their edited volume *Humanistic Geography: Prospects and Problems* in which they raised questions about how we could frame the geographies of the future.

Not surprisingly, those who were interested in questions of experience, meaning, and subjectivity soon expressed concerns about quantitative geography's tendencies to be overly abstract and reductive, as well as to question its focus on certainty. Could a scientific model incorporate subjective experience, perception, attitudes and concepts of social behavior? How could scientific geography represent what increasingly was being seen as the rich, complex, and possibly contradictory forms of lived experience (Ley & Samuels, 1978)? For some, a focus on subjectivity, experience, and behavior led to a dismissive approach to science as being overly abstract and instrumental (Pickles, 1985, 1986). Humanistic geographers recognized the opportunity and the necessity of thinking about human experience outside the domain of cybernetics and systems relationality where humans could be thought about in terms that did not reduce them to abstract rational economic actors. Instead, experience was to be seen either as a form of individual subjectivity or as a complex embedded and over-determined effect that interacted with the world in ways that were not able to be captured within logics of instrumental reason (Lee & Smith, 2004). For others, a focus on behavior and behavioral science mediated these tensions (Downs & Stea, 1973).

Many of these reactions took several forms not typically included in the 'canon' of 'humanistic geography', but it seems important to recognize them as active parallel participants in the 'science/humanities' debates, particularly because of their close allegiances with the humanities. Some were critical—even reactionary—responses on the part of descriptive and regional geographers, most notably Richard

Hartshorne (1939, 1959) who rejected outright the possibility of predictive nomothetic science in what he saw as an idiographic discipline. Equivalent to history, this geography dealt with non-generalizable unique phenomena. There was also skepticism about the reductionism of the 'new scientific geography' from cultural historical geographers such as Carl Sauer, Clarence Glacken, Isiah Bowman, and Peirce Lewis. Their work focused variously on rich traditions of cultural analysis and how cultural landscape variations could be read in multiple ways that required an openness to a wider array of social and cultural sciences, particularly history, anthropology, and archaeology.

In the 1970s, humanistic geographers engaged with the wider humanities. Interest in the histories of maps and representation brought some into contact with archivists, historians, and artists particularly around debates about what constitutes the map, map design, and social uses of maps (e.g., Cosgrove, 1984). For other geographers, the injunction to consider values in geography opened important avenues of welfare and social justice research. For example, David Smith (1994), drawing on political philosophy, particularly Rawlsian theories of justice, rejected the idea that geography could be value-free or morally neutral because spatial relationships involve ethical considerations about justice, inequality, and human welfare (Smith, 1997). Smith drove these ideas into humanistic and radical geography, Gilbert White (e.g., 1969) made parallel arguments about safe water as a human right and water management studies as essential human welfare concerns (see also Kates, 2011), and Birdsall (1996) called for a moral geography founded on regard, respect, and responsibility.

Beyond the Science/Humanism Dichotomy

So far, I have suggested that 'scientific geography' and 'humanistic geography' functioned in parallel with each other, operating in different registers, but with some important common concerns and commitments, particularly about environmental change and social justice. These parallels and common concerns were not immediately apparent in subsequent years as the two sub-fields juggled with their underlying differences in epistemology and how they treated human decision-making. A key example is how spatial analysis was incorporated in the context of computer mapping. For some years these developments submerged its on-going development of spatial analysis into what later became the powerful capacities of GIScience in which, again, key individuals and centers were crucial (such as Michael Goodchild at Santa Barbara and Jack Dangermond at ESRI). One crucial aspect of this parallel development was the way in which each conceived of, and responded to, the notion of 'science'. Here, both groups tended to frame 'science' narrowly, one focusing on a logical positivist (or logical empiricist) conception of science and the other rejecting the same positivism and, by extension, science and scientific method as such (Pickles, 1985).

The project of constructing a universal and unified theory of science fell more broadly to the logical positivists such as Otto Neurath, Rudolf Carnap, Ernst Nagel, Philipp Frank, and Carl Hempel. But, in 1947 Thomas Kuhn's '*The Structure of Scientific Revolutions*' appeared as Volume 2, Issue 2 of the *International Encyclopedia of Unified Science* (contracted by the editors to demonstrate **historically** the progressive nature of science). Despite the editors' initial charge, Kuhn's resulting history of the development of scientific ideas questioned the underlying assumption of the project and its commitment to a particular understanding of how scientific ideas changed. Instead, Kuhn's concepts of paradigm and revolutionary break displayed the history of science as a social, rhetorical, and political practice. These ideas were influential among geographers such as Ron Johnston (1979), who applied the notion of the paradigm to his understanding of the ways in which geography changed through time from one domain to another as ideas were contested and then transformed. Here, the concept of paradigm became an organizational concept used to structure the diverse sub-fields in geography, particularly spatial science and humanistic geography. As distinct paradigms each came to be seen as parallel discursive domains in which research emphasized either analytical/quantitative methods and largely Euclidean geometries of space or qualitative, experiential, and—in some cases—subjective interpretations of space, nature, and lifeworld (see Pickles & Watts, 1992).

Not all scientific and humanistic approaches in geography fell into this 'positivism'/'anti-positivism' framing. For some, the very concept of 'science' as it had consolidated in Anglo-American traditions, with its understanding of science as a universal project of abstraction, analysis, and prediction (whether inductive or deductive), was the problem. The limits of this paradigmatic framing of parallel approaches became clear with the publication in 1959 of an English translation of Marx's early *Economic and Political Manuscripts*. Here both positivist and liberal interpretations of geography were challenged by a humanistic turn in Marxism as social theory and scientific method offered a parallel reading to the more orthodox economistic reading of Marx that had passed down from Moscow Publishers since the earlier part of the century (Peet, 1977). The result was that radical geography quickly became a new pole of political and intellectual activity in the 1970s and 1980s particularly in its engagement with and around the New Left (Pickles & Watts, 1992).

Harvey (1984, 2021) articulated a different kind of explanation, no longer a binary deterministic modeling process, but instead a dialectic relationship between tensions in society that resulted in various transformations of established situations (Sheppard, 2006, 2025). He focused on the city and particularly on the role of capital in restructuring urban form, and was joined by others, such as Gunnar Olsson whose deep understanding of analytics was matched by a recognition that spatial analysis was inherently conservative, reproducing the very conditions it started out describing (see Olsson, 2002); in building predictive models based on existing sets of relationships the geographer and the spatial analyst reproduced those existing relationships.

The wider impact was an intense churning of perspectives and debates about approach, method, and what constituted evidence. Again, key nodes emerged within which debate unfolded, chief among them were the *Newsletter of the Union of Socialist Geographers* where liberal, socialist, Marxist, and anarchist debates flourished and *Antipode: A Journal of Radical Geography* which quickly established itself as the key site for radical geographers to engage with each other on similar grounds of political theory and development studies, and with the equally dynamic feminist critiques of patriarchy, gender, and sexuality. In these debates the geographies attendant on different conceptions of the human being were explored beyond the rational or the liberal individual. Bill Bunge illustrated this mediation of scientific geography and a different humanism when, in 1968 he founded the Society for Human Exploration, followed the next year by the Detroit Expedition (Bascom, 2025).

These new approaches focused on the complexity and uncertainty of human experience but rejected both the rhetorical framing of 'science' against the humanities, the concept of 'science' rooted in the specific reductionist projects of Comtean positivism and logical empiricism, and the liberal individual as the basis for social analysis. Here anti-science rhetoric was replaced by explorations of the cultural dimensions of space, place, and identity, including Cosgrove's (1984) *Social Formation and Symbolic Landscape*, Cosgrove and Daniels (1988) *The Iconography of Landscape*, and James Duncan's (1980) 'The Superorganic in American Cultural Geography' that focused attention on cultural representations, narratives, and symbols in shaping everyday experience.

For others, the dominance of quantitative and positivist approaches during the 1960s and 1970s and the reactions of humanistic geography both overlooked some of the more important political and economic structural dimensions of what it meant to live in this world. Feminist geographers at the time questioned the ways in which abstract modeling of spatial relations and undifferentiated notions of human subjects ignored crucial aspects of the social and political dimensions of space, especially gender. These abstractions overlooked women's experiences and, in effect, made women and other social actors invisible (Massey, 1984; Monk & Hanson, 1982). In response, some feminist geographers focused on subjective, embodied experiences which led to expanded engagement with qualitative and ethnographic methods (Katz & Monk, 1993; Rose, 1993). The result was a series of vibrant geographies that teased out, for example, the important differences between public and private spaces, everyday experiences of domestic spaces, work, housing, and public policy, and the gendered consequences of patriarchal public space (McDowell, 1983; Women and Women and Geography Study Group of the IBG, 1984; Smith, 1989).

By the late 1980s, intersectional perspectives from Black feminism and postcolonial theory, and later non-representational theory (Thrift, 2008), further widened the scope of subjecthood and interdisciplinary engagements, challenging earlier geographies for being too centered on white, middle-class women's experiences (hooks, 1984). Gillian Rose's *Feminism and Geography: The Limits of Geographical Knowledge* (1993) was particularly important in demonstrating how the turn to science and abstract concepts of space constituted an apolitical geography that led to the exclusion of women's experience and lives. She argued that space was more productively seen as embedded in a range of gendered and raced social and cultural practices. This meant that 'other' spaces such as the home, workplace, and street were important objects in which these experiences could be foregrounded and retrieved for a critical geography.

Critical geography flourished from the 1980s on, with many distinct sites emerging for its elaboration. Journals from *Antipode* to *Progress in Human Geography* to Pion's *Environment and Planning D: Society and Space* were key outlets for critical geographers, as well as scholars from other disciplines who were increasingly engaging with geographical ideas and texts. Geographers rapidly expanded their horizons to engage with the wider churning of ideas in social theory, continental philosophy, post-structuralism, and Gramscian inflected cultural theory. Several geography departments became iconic centers of these efforts (e.g., UCLA, Minnesota, Clark, and Penn State among others) and later, one in particular—the University of Kentucky—built one of the stronger interdisciplinary social theory programs in the US; the UK Committee on Social Theory, founded and run jointly by geographer John Paul Jones, Germanist Wolfgang Natter, philosopher Theodore Schatzki and colleagues (the latter two eventually moving their tenure homes into geography) with its journal *disClosure: A Journal of Social Theory*.

In the UK, the Birmingham School of Cultural Studies at the Centre for Contemporary Cultural Studies (CCCS) founded in 1964 by Richard Hoggart was influential for human geographers. Focused on questions of culture, politics, and identity, the CCCS and the writings of Raymond Williams, Stuart Hall, Richard Hoggart, Michael Rustin, and Larry Grossberg defined a new field of critical cultural theory. Grossberg (2010) has referred to this new problem space of culture as a complex and dynamic field where cultural practices, power relations, and social and spatial identities intersect. These intersections enabled new conversations for geographers with British cultural studies with its Gramscian inflection, especially under the influence of Doreen Massey (see 2013 and her other contributions to the New Left journal *Soundings*). As a result, geographers such as Peter Jackson (1989) became much more engaged with questions of indeterminacy, openness, and complexity. Space, place, and environment were, in this sense, evolving and contested across multiple political, economic, and social structures, the outcome of which struggles are never guaranteed (Grossberg et al., 2000). Pickles (2012, 2018) has made this link to geography and the important role played by Doreen Massey, while more recently Jamie Peck (2024) and Eric Sheppard, Helga Leitner, and Jamie Peck (2024) have explicitly linked to cultural studies their call for a conjunctural geographical analysis that examines the complex interplay among societal structures, spatial dynamics, and biophysical processes.

In this expanded interest in culture, hegemony, and interpretation, geographers followed Bernstein in teasing out various key conceptions of science and knowledge through structuralism, hermeneutics, and phenomenology, among other traditions: what Bernstein referred to as *Beyond Objectivism and Relativism* (1983; see also 1978). Nowhere were these ideas clearer than in Derek Gregory's (1979) *Ideology Science and Human Geography* where he explicitly re-read geographical ideas in and through wider debates and perspectives from critical social theory and continental philosophy. Gregory was influenced by Anthony Giddens, and he in turn was influenced by Juergen Habermas who became an important figure for a new radical humanism because of the ways in which he articulated the relationship between science and rationality on the one hand and the question of lifeworld experience on the other. Stressing the centrality of human interest in shaping any claim about knowledge, he emphasized the role of human interest and agency in shaping socio-spatial relations (Habermas, 1972).

The expanded role of translation and publication of continental philosophy in the social sciences had a particularly important role in the geography of the period, perhaps best illustrated by Ed Soja's *Postmodern Geographies* (1989) and Derek Gregory's (1994) *Geographical Imaginations*.

New cohorts of faculty and students steeped in the social moments of the long 1960s strove in the 1980s to come to terms with the science/humanism dichotomies of prior decades. As we will see in the next section, geography provided a particularly open and receptive field for these explorations. In the process, social theory transformed both our understanding of scientific research and humanism, and became a central prerequisite for many, albeit often resisted as jargonistic by more orthodox scholars. At the same time, other areas of geography entered strong relations with trans- and inter-disciplinary networks, linking GIS and environmental geography directly to earth sciences, climate studies, and ecology.

Knowledge Production

Crucial to any understanding of these changing patterns, places, and forms of geographical knowledge production, and one important driver of the proliferation of perspectives that emerged subsequently, was the rapid expansion of universities and the growth in student numbers. For the US, Kavita Pandit (2004) has documented these changes for the period 1948–2001 (Fig. 1). Figure 2 illustrates the growth in AAG membership between 1948 and 2002, when a boom in membership reflected the relaxing of rules on membership, expansion of geography degree programs across the US, and the internationalization of the Annual Conference of the AAG (Fig. 3).

In this process of numerical expansion, the field also became more diverse. While the 1950s and 1960s appear to have seen a decline in post-war enrolments of women in geography (a combination of post-war declines in opportunity, the expansion of male geographers in teaching, and the increasing centrality of quantitative approaches in the curriculum), their numbers rebounded in the 1970s (Fig. 3). More aspects of geographical education incorporated questions that these changing gendered cohorts brought to their departments.

New developments in philosophy, social theory, sociology and political science, along with the emergence of interdisciplinary studies in science and the humanities, drew students to take courses outside of departments of geography as well as their

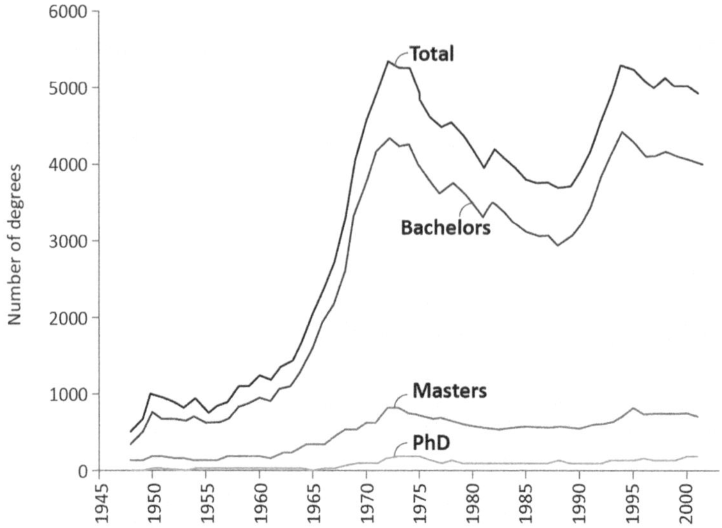

Fig. 1 Geography degrees awarded, 1948–2001 (Modified from Pandit, 2004)

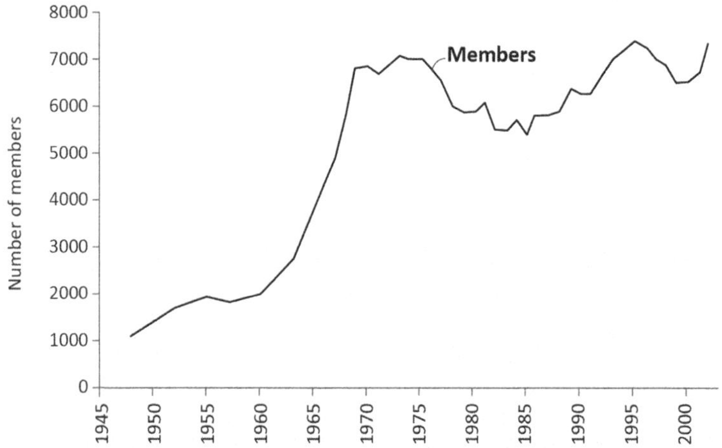

Fig. 2 Trends in AAG membership, 1948–2002 (Modified from Pandit, 2004)

topical and thematic and methodological courses inside their departments, and this too had a geography to it. At public, land-grant universities including Berkeley, British Columbia, Kentucky, Minnesota, Ohio State, Penn State, Simon Fraser, UCLA, and Washington among others, diverse constellations of faculty members were assembled, often by senior scholars who recognized the opportunities for disciplinary expansion. A smaller number of US private universities also became key nodes for innovation: at Clark University, Susan Hanson, Dick Peet, David Stea, Ben Wisner, and others were instrumental in expanding the range of ideas within

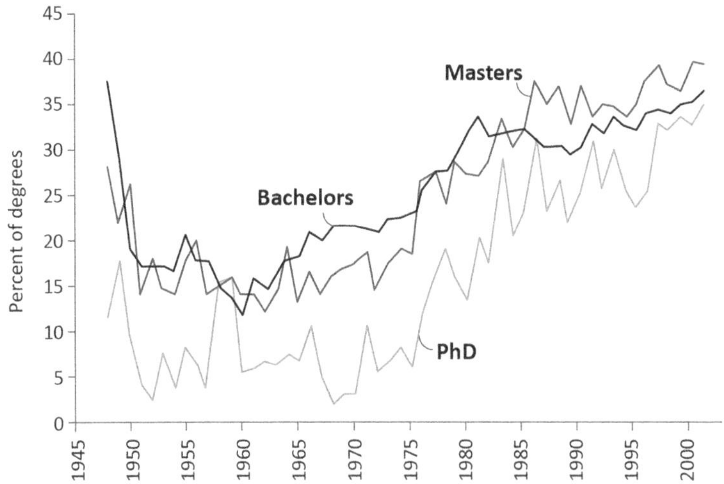

Fig. 3 Percent of geography degrees awarded to women, 1948–2001 (Modified from Pandit, 2004)

radical geography, while at Johns Hopkins University David Harvey gathered a group of critical scholars working on Marxism and geography, among them Dick Walker, Neil Smith, Cindi Katz, Patrick Bond, and Eric Swyngedouw.

At the same time as these programs opened the discipline, publishing houses and federal and private funding agencies were underwriting strong inter-disciplinary research, in turn increasing the number and diversity of publishing outlets for new perspectives in areas such as sustainability, environmental awareness, geopolitics, planetary science, and race/ethnicity through new journals such as *Geopolitics, Global Environmental Change, Dialogues in Human Geography, Dialogues in Physical Geography, Global Networks, Territories-Politics-Government, Mobilities* and the new radical journal *Human Geography.*

At these key nodes, and quickly at others, groups of critical geographers emerged variously engaged with Marxism, anarchism, feminism, phenomenology, and nearly always reading out from the field into wider contexts of contemporary political theory and philosophy. Not surprisingly, these nodal groups quickly forged their own social and intellectual networks at regional and national conferences, and later became active networks on journal editorial boards, including new journals where geographers such as Nigel Thrift, Dick Peet, and Eric Sheppard and editors such as Peter Wissoker at Guilford and John Davey at Blackwell, were particularly influential. Derek Gregory wrote in his tribute to Davey on his passing at the age of 72, "[He] was one of those rare publishers who believed passionately that books created their audiences and that geography was so much more than a textbook machine. He didn't spurn textbooks, but he had a non-mercenary and thoroughly ambitious sense of what they ought to strive for." (https://www.aag.org/memorial/john-charles-armstrong-davey/) Each major journal passed to subject editors who supported these changes. Stan Brunn at the *Annals of the Association of American Geographers* opened the journal to a wider range of topics and perspectives, a practice expanded by John Paul Jones when he later became editor of the *Annals* and encouraged diverse ideas, formats, and papers from non-geographers. In Britain, parallel developments were taking place in the *Transactions of the Institute of British Geographers* and new kinds of journals emerged such as the Pion Environment and Planning Series particularly *Environment and Planning D: Society and Space* where from 1983 onward Nigel Thrift played a central editorial role in diversifying the field. Other journals such as *Geoforum*, *GeoJournal*, and Dick Peet's online journal *Human Geography* were among a group of new outlets supporting diversity of regional, language, disciplinary, and inter-disciplinary origin. Other institutional supports such as funding agencies also became important in developing new approaches to research. The National Science Foundation, particularly under the leadership of Ron Adler, Tom Baerwald, and their colleagues, began to expand the scope of funding to include more diverse approaches to research while also expanding the levels of funding for spatial analysis across the discipline (see Pickles, 1982).

Conclusion

Quantitative methods and positive science opened new avenues for practical geographies of locational analysis, consumer modeling, urban planning, and regional economic development. At the same time, these approaches became focal points of concern about positivism, reductionism, and technicism as humanistic geographers claimed the centrality of human experience beyond the rational economic optimizer or satisficer individual.

The intervening decades have both consolidated these distinctions and legitimized them under the rubric of paradigm and sub-disciplinary specialization. But, at the same time, the intervening decades have seen geography departments, in a sense, settle down from the experience of 'science wars'. The ontological and epistemological questions about the humanities and science have slipped into the background, if they hold any purchase at all. Analytical and empirical approaches to research have become well established, even reinvigorated by advances in GIScience. In many departments, these subfields have become embedded in 'big science' and large funding benefits have accrued not only to specific analytical areas of inquiry, but also to scholars, research groups, and department capacities more widely. The proliferation of methodologies around qualitative research methods, identity politics, and critical approaches to research have also contributed programmatically to recruiting a wider range of skilled students into the discipline. Today, US departments tend to hire clusters around areas of strength, balancing science and humanities, and expanding the faculty mix to include scholars trained in related disciplines such as ecology, environmental science, engineering, anthropology, history, and other areas of the humanities. Few modern departments seek to cover the entire geographic 'waterfront' and the most successful of them seem to be those that strive to maintain diverse epistemological and methodological skill sets among their faculty and students, while maintaining extensive inter-disciplinary relations with other fields and scholars from other countries.

C.P. Snow's (1959) *Two Cultures* may remain an active part of many geographers' outlook and teaching, but generally contemporary geography seems more open to difference and critical dialogue. In the process, the question of political ecology, human ecology, the capitalocene, and non-human ecologies have, variously, reworked many of the assumptions that underpinned humanistic and scientific geographies of human subjectivity and an external physical world. As Anna Tsing suggests, "human nature is an interspecies relationship" (Kirksey et al., 2014, p. 2) and in exploring that relationship after the 1980s many exciting geographies have been, and are still to be, written (see Luisetti, 2023).

Acknowledgements Thanks to Stan Brunn for his thoughtful suggestions to ground this chapter in concrete institutions and practices, and for his insightful suggestions on the draft manuscript.

References

Abler, R. J., Adams, J. S., & Gould, P. (1971). *Spatial organization: The geographer's view of the world*. Prentice-Hall.

Ackerman, E. A. (1954). Geographic training, wartime research, and immediate professional objectives. *Annals of the Association of American Geographers, 44*(3), 265–276.

Barnes, T. J. (2025). Abstract. Keeping human and physical geography together: Richard Chorley and Peter Haggett's scientific turn. In N. Castree, T. Barnes, & J. Salmond (Eds.), *Making geography matter: The past and present of a changing discipline*. Routledge. https://www.taylorfrancis.com/chapters/edit/10.4324/9781003343240-5/keeping-human-physical-geography-together-trevor-barnes

Barnes, T. J., & Bergmann, L. R. (2022). The early years: William Bunge and theoretical geography. In F. Gyuris, B. Michel, & K. Paulus (Eds.), *Recalibrating the quantitative revolution in geography*. Routledge.

Barnes, T. J., & Wilson, M. W. (2014). Big data, social physics, and spatial analysis: The early years. *Big Data & Society, 1*(1). https://doi.org/10.1177/2053951714535365

Bascom, J. (2025). The historical and philosophical emergence of radical geography. *The Geographical Bulletin, 22*(1), 7–13. https://digitalcommons.kennesaw.edu/thegeographicalbulletin/vol22/iss1/2

Bernstein, R. (1978). *The restructuring of social and political theory*. University of Pennsylvania Press.

Bernstein, R. (1983). *Beyond objectivism and relativism: Science, hermeneutics, and praxis*. University of Pennsylvania Press.

Berry, B. J. L., & Marble, D. F. (Eds.). (1968). *Spatial analysis: A reader in statistical geography*. Prentice-Hall.

Birdsall, S. S. (1996). Regard, respect and responsibility: Sketches for a moral geography of the everyday. *Annals, Association of American Geographers, 86*, 619–629.

Boulding, K. E. (1956). *The image: Knowledge in life and society*. University of Michigan Press.

Boulding, K. E. (1962a). *Conflict and defense: A general theory*. Harper.

Boulding, K. E. (1962b). Economic development as an evolutionary system. *American Economic Review, 52*(2), 42–45.

Browning, C. (1982). *Conversations with geographers: Career pathways and research styles*. Department of Geography, University of North Carolina Studies in Geography No. 16.

Brunn, S. D. (2004). The midwesternization of southern geography departments. In J. O. Wheeler & S. D. Brunn (Eds.), *The role of the south in the making of American geography: Centennial of the AAG 2004*. Bellwether for the Southeastern Division, Association of American Geographers.
Bunge, W. (1962). *Theoretical geography*. The Royal University of Lund Department of Geography. Lund Studies in Geography.
Bunge, W. (1979). Perspective on theoretical geography. *Annals of the Association of American Geographers, 69*, 169–174.
Buttimer, A. (1974). *Values in geography*. Association of American Geographers. Commission on College Geography. Resource Paper 24.
Chorley, R. J., & Haggett, P. (1967). *Models in geography*. Methuen.
Christaller, W. (1933). *Die Zentralen Orte in Sűddeutschland*. Verlag von Gustav Fischer, Jena (transl. Central Places in Southern Germany, 1966).
Cosgrove, D. (1984). *Social formation and symbolic landscape*. University of Wisconsin Press.
Cosgrove, D., & Daniels, S. (Eds.). (1988). *The iconography of landscape: Essays on the symbolic representation, design and use of past environments*. Cambridge University Press.
Dickinson, R. E. (1976). *The regional concept: The Anglo-American leaders*. Routledge and Kegan Paul.
Downs, R. M., & Stea, D. (1973). *Image and environment: Cognitive mapping and spatial behavior*. Routledge.
Duncan, J. S. (1980). The superorganic in American cultural geography. *Annals of the American Association of Geographers, 70*(2), 181–198.
Ferretti, F. (2020). History and philosophy of geography II: Rediscovering individuals, fostering interdisciplinarity and renegotiating the 'margins'. *Progress in Human Geography, 45*(4), 890–901.
Galison, P. (1994). The ontology of the enemy: Norbert Wiener and the cybernetic vision. *Critical Inquiry, 21*(1), 228–266.
Garrison, W. L., & Berry, B. J. L. (1958). Alternate explanations of urban rank-size relationships. *Annals of the Association of American Geographers, 48*(1), 83–91.
Gould, P. (1969). *Spatial diffusion*. Commission on College Geography Resource Paper No. 4. Association of American Geographers.
Gould, P. (1975). People in information space: The mental maps and information surfaces of Sweden. *Lund studies in geography series B: Human Geography* No. 42.
Gould, P., & Pitts, F. R. (2002). *Geographical voices: Fourteen autobiographical essays*. Syracuse University Press.
Gould, P., & White, R. (1974). *Mental maps*. Penguin Books.
Gregory, D. (1979). *Ideology, science and human geography*. St. Martin's Press.
Gregory, D. (1994). *Geographical imaginations*. Blackwell.
Grossberg, L. (2010). *Cultural studies in the future tense*. Duke University Press.
Grossberg, L., McRobbie, A., & Gilroy, P. (Eds.). (2000). *Without guarantees: In honour of Stuart Hall*. Verso.
Habermas, J. (1972). *Knowledge and human interests*. Beacon Press.
Hägerstrand, T. (1953). Innovation diffusion as a spatial process. *Lund Studies in Geography, Series B*, No. 4.
Haggett, P. (1965). *Locational analysis in human geography*. Edward Arnold.
Hartshorne, R. (1939). *The nature of geography: A critical survey of current thought in the light of the past*. Association of American Geographers.
Hartshorne, R. (1959). *Perspective on the nature of geography*. Association of American Geographers.
Harvey, D. (1969). *Explanation in geography*. Edward Arnold.
Harvey, D. (1984). On the history and present condition of geography: An historical materialist manifesto. *The Professional Geographer, 36*(1), 1–11.

Harvey, D. (2021). Reflections on an academic life. *Human Geography, 15*(1). https://doi.org/10.1177/19427786211046291
hooks, b. (1984). *Feminist theory: From margin to center*. South End Press.
Huxley, A. (1932). *Brave new world*. Chatto and Windus.
Isard, W. (1956). *Location and space-economy: A general theory relating to industrial location, market areas, land use, trade and urban structure*. MIT Press.
Isserman, A. (2004). Intellectual leaders of regional science: A half-century citation study. *Papers in Regional Science, 83*(1), 91–126.
Jackson, P. (1989). *Maps of meaning: An introduction to cultural geography*. Unwin Hyman.
Jacobs, J. (1992 [1961]). *The death and life of great American cities*. Vintage.
Jacobs, J. (1969). *The economy of cities*. Vintage.
Johnston, R. J. (1979). *Geography and geographers: Anglo-American human geography since 1945*. Wiley.
Kates, R. W. (2011). *Gilbert F. White, 1911-2006, A biographical memoir*. The National Academies.
Katz, C., & Monk, J. (Eds.). (1993). *Full circles: Geographies of women over the life course*. Routledge: International Studies of Women & Place Series.
Kirksey, E., Schuetze, C., & Heimreich, S. (2014). *The multispecies salon*. Duke University Press.
Kurokawa, K. (1988). *Kisho Kurokawa: The architecture of Symbiosis*. Rizzoli International Publications.
Leatham, T. (2023, 27 January). The story behind the song: Malvina Reynolds and Pete Seeger's housing crisis satire 'Little Boxes'. *Far Out Magazine*. https://faroutmagazine.co.uk/story-behind-pete-seeger-little-boxes/
Lee, R., & Smith, D. M. (Eds.). (2004). *Geographies and moralities*. Blackwell.
Lefebvre, H. (2022). *On the rural: Economy, sociology, geography*. University of Minnesota Press.
Ley, D., & Samuels, M. (Eds.). (1978). *Humanistic geography: Prospects and problems*. Croom Press.
Lösch, A. (1940). *Die Räumliche Ordnung Der Wirtschaft*, Jena: Gustav Fischer. Translated from German by W. H. Woglom (1954). The economics of location. Yale University Press.
Luisetti, F. (2023). *Non-human subjects: An ecology of earth-beings*. Cambridge University Press.
Massey, D. (1984). *Spatial divisions of labour: Social structures and the geography of production*. Macmillan.
Massey, D. (2013). Vocabularies of the economy. *Soundings, 54*.
McDowell, L. (1983). Towards an understanding of the gender division of urban space. *Environment and Planning D: Society and Space, 1*(1), 59–72.
Mead, M. (1968). Cybernetics of cybernetics. In H. von Foerster, J. D. White, & J. K. Russell (Eds.), *Purposive systems* (pp. 1–11). Spartan Books.
Monk, J., & Hanson, S. (1982). On not excluding half of the human in human geography. *The Professional Geographer., 34*(1), 11–23.
Oakeshott, M. (1991 [1962]). *Rationalism in politics and other essays*. Liberty Fund.
Olsson, G. (1965). *Distance and human interaction: A review and bibliography*. Regional Science Research Institute.
Olsson, G. (1968). *Distance, human interaction and stochastic processes: Essays on geographic model building*. Department of Geography, University of Michigan.
Olsson, G. (1975). *Birds in egg*. Michigan Geographical Publications, No. 15.
Olsson, G. (2002). Glimpses. In P. Gould & F. R. Pitts (Eds.), *Geographical voices: Fourteen autobiographical essays* (pp. 237–268). Syracuse University Press.
Orwell, G. (1949). *1984*. Secker and Warburg.
Pandit, K. (2004). Geography's human resources over the past half-century. *The Professional Geographer, 56*(1), 12–21.
Peck, J. (2024). Articulating conjunctural analysis. *Dialogues in Human Geography, 14*(3), 504–508.
Peet, R. (1977). *Radical geography: Alternative viewpoints on contemporary social issues*. Maaroufa Press.

Pickles, J. (1982). 'Science' and the funding of human geography. *The Professional Geographer, 34*(4), 387–392.
Pickles, J. (1985). *Phenomenology, science, and geography: Spatiality and the human sciences.* Cambridge University Press.
Pickles, J. (1986). *Geography and humanism (CATMOG).* GeoBooks.
Pickles, J. (2012). The cultural turn and the conjunctural economy: Economic geography, anthropology, and cultural studies. In T. Barnes, J. Peck, & E. Sheppard (Eds.), *The Wiley-Blackwell companion to economic geography* (pp. 537–551). Wiley-Blackwell.
Pickles, J. (2018). From the "the" north to "the" south: Spatializing the conjuncture in British cultural studies. In M. Werner, J. Peck, R. Lave, B. Christophers, et al. (Eds.), *Doreen Massey critical dialogues* (pp. 137–149). Agenda Publishing.
Pickles, J., & Watts, M. J. (1992). Paradigms for inquiry? Criticism and explanation in contemporary human geography. In R. A. Abler, M. Marcus, & J. Olson (Eds.), *Geographies inner worlds. Pervasive themes in contemporary American geography* (pp. 301–326). Rutgers University Press.
Ray, D. M. (1972). Review: Ronald Abler, John S. Adams and Peter Gould (1971) *Spatial organization: The geographer's view of the world.* Englewood Cliffs, NJ: Prentice-Hall. *Economic Geography, 48*(4), 447–448.
Relph, E. (1976). *Place and placelessness.* Pion.
Rose, G. (1993). *Feminism and geography: The limits of geographical knowledge.* Polity Press.
Schaefer, F. K. (1953). Exceptionalism in geography: A methodological examination. *Annals of the Association of American Geographers, 43*(2), 226–249. https://doi.org/10.1080/00045605309352140
Sheppard, E. (2006). David Harvey and dialectical space-time. In N. Castree & D. Gregory (Eds.), *David Harvey: A critical reader* (pp. 121–141). Wiley.
Sheppard, E. (2025). Radicalizing geography: The case of David Harvey's Marxism. In N. Castree, T. Barnes, & J. Salmond (Eds.), *Making geography matter: The past and present of a changing discipline.* Routledge.
Sheppard, E., Leitner, H., & Peck, J. (2024). Conjunctural analysis. In D. Richardson (Ed.), *The international encyclopedia of geography: People, the earth, environment, and technology* (pp. 1–7). Wiley.
Smith, D. M. (1994). *Geography and social justice.* Blackwell.
Smith, D. M. (1997). Geography and ethics: A moral turn? *Progress in Human Geography, 21*(4), 583–590.
Smith, S. J. (1989). Women, work and housing: The experience of self-employed women in West Yorkshire. In N. Thrift & R. Peet (Eds.), *New models in geography* (Vol. 1, pp. 232–247). Unwin Hyman.
Snow, C. P. (1959). *The two cultures. The Rede lecture.* Cambridge University Press.
Soja, E. W. (1989). *Postmodern geographies: The reassertion of space in critical social theory.* Verso.
Stewart, J. Q. (1941). On distance and social influence. *Science, 93*(2404), 89–90.
Stewart, J. Q. (1947). Empirical mathematical rules concerning the distribution and equilibrium of population. *Geographical Review, 37*(3), 461–485.
Stewart, J. Q. (1948). Demographic gravitation: Evidence and applications. *Sociometry, 11*(1/2), 31–58.
Stewart, J. Q. (1950). The development of social physics. *American Journal of Physics, 18*(4), 239–253.
Stewart, J. Q., & Warntz, W. (1958). Physics of population distribution. *Journal of Regional Science, 1*(1), 99–126.
Strahler, A. N. (1952). Dynamic basis of geomorphology. *Geological Society of America Bulletin, 63*(9), 923–938.
Thrift, N. (2008). *Non-representational theory: Space, politics, affect.* Routledge.

Tuan, Y.-F. (1974). *Topophilia: A study of environmental perception, attitudes, and values*. Prentice-Hall.
Tuan, Y.-F. (1977). *Space and place: The perspective of experience*. University of Minnesota Press.
von Foerster, H., Mead, M., & Teuber, H. L. (Eds.). (1951). *Cybernetics: Circular casual and feedback mechanisms in biological and feedback mechanisms in biological and social systems*. Transactions of the seventh conference. Josiah Macy, Jr. Foundation.
Warntz, W. (1957). Geography of prices and spatial interaction. *Papers in Regional Science, 3*(1), 118–129.
White, G. F. (Ed.). (1969). *Water, health and society, selected papers by Abel Wolman*. Indiana University Press.
Wiener, N. (1948). *Cybernetics: Or control and communication in the animal and the machine*. MIT Press.
Wiener, N. (1950). *The human use of human beings*. Houghton-Mifflin and the Riverside Press.
Women and Geography Study Group of the IBG. (1984). *Geography and gender: An introduction to feminist geography*. Hutchinson.
Zamyatin, Y. (1924). *We*. E. P. Dutton.

Settlement of the Amish Amongst the "English"

John A. Cross

The first Amish settlements in North America were established in eastern Pennsylvania nearly three centuries ago (Crowley, 1978). Since then, the Amish population has grown to nearly 385,000, with 617 settlements in the United States and 20 settlements in Canada (Young Center, 2023a). Within the United States, Amish settlements are found in 32 states, with four states—Missouri, Ohio, Pennsylvania, and Wisconsin—each having over 60 settlements. Four Canadian provinces had Amish settlements in 2023 (Young Center, 2023a). The places where the Amish settlements were established in the eighteenth and early nineteenth centuries, such as in parts of Pennsylvania and Ohio, have long witnessed expanding Amish populations with a history of interactions with surrounding populations. Other locations, such as Wisconsin, Missouri, Kentucky, and New York, have seen almost all of their Amish settlements being established over the past half century. In both types of places the Amish, whose identity and religious beliefs separate them from modern society, are occupying places in which they are either surrounded by other populations, or where their farmsteads are intermixed on the landscape among others. Thus, the Amish provide a wonderful example to include in a collection of chapters that focus upon *Parallel Universes: Places Where People Live Together and Separately.*

This chapter first briefly discusses various attributes of the Amish that distinguish and separate them from the surrounding population. It then explores the locations of Amish settlements and notes continuing changes in the spatial distribution of their settlements. Then it considers the spatial distribution of Amish landholdings within their settlements, demonstrating the close proximity of many Amish households to the non-Amish, or "the English" population as the Amish would describe

J. A. Cross (✉)
Department of Geography, University of Wisconsin, Oshkosh, WI, USA
e-mail: cross@uwosh.edu

S. D. Brunn (ed.), *Transdisciplinary Research Across Knowledge Worlds*,
https://doi.org/10.1007/978-3-032-19491-6_3

them. Finally, the chapter provides examples illustrating the constrained nature of social and economic interactions between the Amish and their neighbors.

Distinguishing Characteristics of the Amish

The Amish are a devout Anabaptist group following the teachings of Jakob Ammann in the German-speaking area of Switzerland, separating the Amish from the Mennonites on doctrinal grounds. Facing severe persecution from both Roman Catholics and other Protestant groups, many Amish first fled down the Rhine River into the Palatinate in the late seventeenth century. Many subsequently sought refuge in colonial Pennsylvania, which had gained a reputation for attracting not only Quakers, but many other Pietist groups, including the Mennonites. A second wave brought most of the remaining Amish in Europe to the United States in the first half of the nineteenth century. Europe lost its last Amish inhabitants nearly a century ago. Given that the Amish do not proselytize and their number of converts is exceedingly small (Kraybill et al., 2013), almost all of the Amish share the same Swiss–German ancestry.

Following the teachings of Menno Simons and Jakob Ammann, the Amish are pacifist Anabaptists, not believing in infant baptism. Their religious beliefs, focused upon the *New Testament*, embrace both discipleship and obedience, shown by separation from others and avoiding the temptations of modernity (Kraybill et al., 2013; Kraybill & Olshan, 1994). The Amish worship together as a congregation in a geographically delineated church district. As a covenant community lacking a hierarchal structure, the congregation and its lay pastors and bishop interpret the Scriptures and standards of behavior. Their code of conduct is governed by the *Ordnung*, an often unwritten set of guidelines spelling out a wide variety of regulations that govern daily life. It is interpreted by the local bishop, who typically oversees two congregations. As Kraybill (2010, p. 161) explains, the *Ordnung* goes beyond behaviors "that are not directly addressed by Scripture" and encompasses "issues related to moral behavior, dress, participation in the larger society, and selective ownership of technology."

Church districts of the same affiliation typically follow a similar *Ordnung*, and local bishops often discuss issues with other bishops. Thus, while the interpretation of the *Ordnung* is relatively consistent within Amish settlements sharing the same affiliation—such as that of Lancaster County which not only governs most of the Amish within its namesake Pennsylvania county, but 31 Amish settlements in ten states (Stoltzfus, 2022)—variations exist among settlements considered as having different affiliations (Petrovich, 2017). For many behaviors, Amish of all affiliations have similar restrictions (Fig. 1), such as no radio or television, use of horse and buggies for everyday transportation, having no religious statuary or "graven" images—including paintings, drawings, or photographs displaying the human face—and never flying a flag within the farmyard—so different from many of their non-Amish neighbors. Visible differences among Amish of various affiliations may

Fig. 1 Amish buggy along road south of Intercourse, Lancaster County, Pennsylvania. Note the clothes on the line of the Amish farmstead

be displayed regarding the color of their horse-drawn buggy—should it be black or gray, or whether it should display red reflectors or lights, and if so what type of lights or reflectors. Likewise, different settlements utilize open versus enclosed buggies, and those whose buggies have tops may vary in its configuration and color, ranging from whites or yellow to grey to black. Likewise, subtle differences in costume exist.

The Amish do not erect church buildings. Instead, the Amish living within a church district, typically twenty to thirty families, worship every-other-week in the house, basement, workshop, or barn of a member family. A visible indication of the location of the worship service is the church wagon, packed with benches and hymnals for use during the three-hour worship service, parked at the farmstead of that week's host family. On the Sunday of worship twenty or more buggies may be parked in the host's farmyard and along the road. The Amish have their own cemeteries, with graves indicated by simple markers rather than crosses.

Amish children who remain within the faith, as about 80–95% do—depending upon affiliation (Nolt, 2015), invariable marry other Amish persons, typically those who are of the same or a closely related affiliation. Many of the Amish families are closely related, with dozens of cousins living nearby. Amish men and women usually marry in their early twenties and have large-sized families, with the typical married woman giving birth to seven or eight children (Donnermeyer, 2023). The Amish often live in multi-generational homes, including grandparents and unmarried siblings. However, given their large number of children, the median age within their settlements is half that of the nation as a whole. For example, the median age is 19 years for the Elkhart-LaGrange Amish settlement in Indiana (Meyers, 2022,

p. 67). Amish settlements display higher population densities than do other rural communities, a result of their higher fertility rate, greater family size, and much smaller farms. Thus, one sees far more children and adults in the rural landscape of Amish settlements.

The use of the Pennsylvania Dutch (Deitsch) or Pennsylvania German language is widespread amongst the Amish, with the exception of several Indiana settlements where Bernese Swiss German is spoken. Not only is Pennsylvania Dutch the everyday language of the home, but it is often used in conversation among Amish individuals within their settlements and those visiting from other Amish settlements. An archaic version of German is used in worship, both in the hymns and scriptural readings, and Amish students are taught "to read German in old fraktur script" (Kraybill et al., 2013, p. 123). Nevertheless, the Amish are also fluent in English, which most speak without an accent, being the language of instruction at school. Amish children are typically educated through the eighth grade in Amish parochial schools, although in a few locations, such as in Holmes County, Ohio, some of the Amish children attend public schools (Hurst & McConnell, 2010). Until the mid-1950s many Amish children attended one-room public schools, yet their parents resisted the consolidation of those schools into larger institutions, thus eliminating interactions with non-Amish schoolmates. Today, Amish parochial school students are taught by an Amish adult whose formal education likewise ended in the eighth grade. Nevertheless, the typical Amish eighth grade student compares quite favorably in their mastery of mathematics and reading relative to students educated in public schools, but they receive far less instruction in science, literature, and social studies (Kraybill et al., 2013). Following their formal schooling, many Amish youth spend several subsequent years in informal apprenticeships. Thus, both the segregation of their children into Amish-only schoolhouses and frequent usage of a language different from that used by their "English" neighbors helps maintain their distinct identity.

Amish men, women, and children are readily identified by their distinctive appearance. Their clothing is plain and modest, typically black or solid dark colors. The trousers of the men are held in place by suspenders rather than belts, and they wear a black vest over their shirt. Women wear long dresses, likewise of dark colors. No zippers are used. Both men and women typically cover their heads. Shorts are not worn. Women wear bonnets, which they call kapps, which are dark blue or black if unmarried, and white thereafter. Men who are married have long beards, but no mustaches. Men, both married as well as young boys, wear straw hats with wide brims. Neither men nor women wear any jewelry, and among many, not even wristwatches. Piercings and tattoos are never seen among the Amish, in contrast to many among the "English" population. Typical Amish clothing can be seen drying on an outdoor clothesline.

Religious prohibitions regarding a variety of technologies conspicuously distinguish the Amish from their non-Amish neighbors. The use of horse-drawn buggies and wagons, which contribute the road droppings that frequently raise the ire of the "English" in those communities that they visit for shopping, and the posting of highway signs warning motorists about the presence of horse-drawn vehicles, denote the existence of an Amish community (or one of the Groffdale Conference of Old Order Mennonites, whose numbers are much smaller than the Amish). Likewise, within the fields of the Amish (Fig. 2) one sees horse-drawn plows, harrows, and other types of horse-drawn equipment—even mechanical harvesters pulled by horses. Avoidance of tractors in the fields is less a religious concern about the technology than a concern about its displacement of labor needed upon the farm. Given Amish beliefs that working in the fields is spiritually inspired (Ericksen et al., 1980; Hostetler, 1987), use of equipment that unduly displaces farm workers is avoided. Depending upon the settlement, bicycles may or may not be permitted, and there are communities that only permit bicycles that function more like foot scooters.

The Amish generally prohibit the use of electricity from the grid or electric utility, thus one does not see power lines connecting to the farmstead, even if a local utility line parallels the road in front of the farmstead. Usage of electricity is tightly controlled within many Amish settlements, but there is a wide variation of what is acceptable by the local *Ordnung*. In some settlements, one will see virtually no electric usage, except for possible battery-powered flashlights. In other settlements, Amish farmsteads may feature solar panels or wind turbines to power a wide variety of equipment and lighting. In some Amish settlements, farmers have gasoline powered generators to operate equipment in their barns or workshops, while in other settlements such equipment is operated using belts attached to stationary tractors. Radio and television are proscribed in all Amish settlements, thus the lack of a television antennae or satellite dish distinguishes Amish homes from many of their

Fig. 2 Team of horses pulling seed planter on Amish farm south of Intercourse, Lancaster County, Pennsylvania

rural neighbors. The use of a telephone is highly dependent upon the local *Ordnung*, ranging from outright prohibition to permitting a phone to be located in a booth within the farmyard or its placement in a barn or workshop (Umble, 1994). Less conspicuous cell phones have been acquired by the Amish in some communities, although business usage is considered more appropriate than personal (Granville & Gilbertson, 2017).

Farming has long been a favored occupation among the Amish (Fig. 3). Indeed, "farming [is] held in higher regard than any other occupation" (Hurst & McConnell, 2010, p. 187). Even though many Amish men in the larger settlements are increasingly employed in non-farm industries, they still need barns to shelter horses and store hay and small fields to graze their animals and plant a kitchen garden. Thus, most Amish homesites conjure the appearance of a farmstead. Recent data for the Greater Lancaster County settlement indicates that 41.0% of Amish adult males are engaged in agriculture, 29.4% work as carpenters, 0.6% work in sawmills, and 29.0% non-farm occupations (Donnermeyer, 2023, p. 9). Within Indiana's Elkhart-LaGrange Amish settlement, which has been most impacted by its population growth exceeding its land base, only 15% of its adult males work in farming, while 55% work in factories, 19% in shops, 5% as carpenters, and 7% in miscellaneous other vocations (Meyers, 2022, p. 72). Elsewhere, such as in Wisconsin, farming remains prominent, employing 58% of the households, with other types of rural employment often taking place on Amish farmsteads, including sawmill operations, woodworking, and carpentry (Cross, 2018, 2023). Among those Amish employed in farming, large numbers in Wisconsin, Minnesota, Michigan, Ohio, and Pennsylvania are engaged in dairying, although their numbers have been declining, such as within Wisconsin (Cross, 2021, 2023).

Amish farmsteads are visually distinctive from those of their "English" neighbors (Cross, 2016; Noble, 1986). Many of the farmhouses, even in locations where new Amish settlements are being established, have been constructed by the Amish,

Fig. 3 Amish dairy farm and farmstead with barn and numerous farm buildings northeast of Berlin, Ohio, within the Holmes County Amish settlement

Fig. 4 White painted wood frame Amish farmhouse in the Town of Tomah, Monroe County, Wisconsin. Note the windmill, a feature seen on many Amish farmsteads

either replacing old run-down structures or adding new farmhouses when the Amish subdivide existing farms that they purchase. These houses (Fig. 4) are typically large two-story plain wood frame structures, which are blocky and often painted white. In the most conservative Amish communities, outhouses are visible. One or more black buggies are present, and work horses may be grazing. Amish farmsteads are far more likely to have working windmills, and extensive kitchen gardens are seen more frequently than upon non-Amish farms. The fields also hint at the presence of the Amish. Even if the horse-drawn plow is not at work in the field, the field that is so cultivated often displays rows that are not as straight or evenly spaced. Following the cutting of the hay or harvesting the grain—be that corn or wheat—stacks of shocked grain dot the field (Fig. 5). Many Amish farmsteads include workshops that employ some of the family, such as woodworking. Signs may be present advertising farm produce, greenhouse nursery plants, candies, furniture and other wood products, quilts, or leather goods for sale, clearly noting "No Sunday Sales."

Amish farmsteads can often be identified by the absence of items frequently seen on "English" properties. Flags are never displayed. Political yard signs are not exhibited, even during election season. Religious statuary, so often seen among certain ethnic populations, is never seen. Power lines do not run to the house. Television antennae and satellite dishes are always absent. Lightning rods are missing from house and barn roofs.

Fig. 5 Rows of shocked wheat in a field with an Amish farmstead in the distance. Photo taken in the Cashton Amish settlement within Monroe County, Wisconsin

Amish Settlement Distribution

Today, the distribution of the Amish population has expanded from their traditional focus upon Pennsylvania and the states that evolved from the eighteenth century's Northwest Territory. The United States contains over 96% of all Amish settlements, as shown in Fig. 6. To the west, there are Amish settlements in Montana and eastern Idaho. To the east there are Amish settlements in Maine and in the Canadian Maritime provinces of New Brunswick and Prince Edward Island. To the north Amish settlements are located in Manitoba and Ontario, while the southernmost North American settlement is located in Florida. During the past decade, two settlement attempts were made in South America, with the one in Bolivia persisting.

The two largest Amish settlements are found in Ohio and Pennsylvania. Pennsylvania's Lancaster County area settlement, the nation's oldest, dates from 1757, although an earlier, now extinct, settlement was established in Berks County, Pennsylvania in 1717 (Crowley, 1978). The Lancaster County settlement spread into adjacent Chester and Dauphin Counties of Pennsylvania and Cecil County, Maryland. Its estimated population was 44,315 persons in 2023, who were accommodated by 257 church districts (Donnermeyer, 2023; Young Center, 2023b). It has the largest population and the second greatest number of church districts of any Amish settlement. The Greater Holmes County, Ohio Amish settlement, founded in 1808, has now extended across portions of five counties. It had the second largest population, 39,525 persons, but the greatest number of church districts, 317

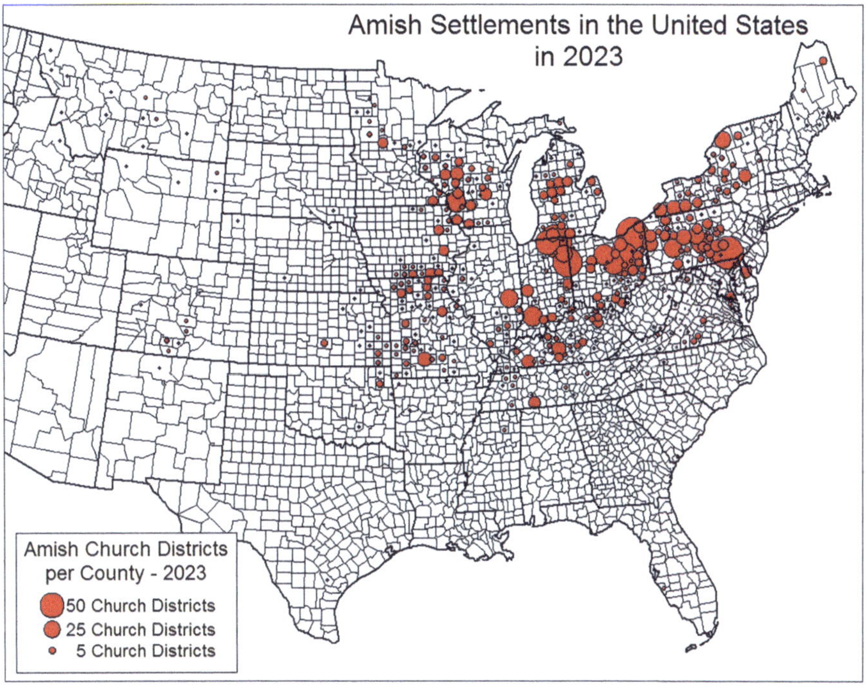

Fig. 6 Distribution of Amish settlements across the United States in 2023, with their sizes indicated by the number of Amish church districts (congregations) per county. Data from Young Center (2023c)

(Donnermeyer, 2023; Young Center, 2023c). The Elkhart–LaGrange Counties Amish settlement, established in 1841 and now also including part of adjacent Noble County in northeastern Indiana, is the nation's third largest, with 215 church districts that include 28,275 persons. The fourth largest settlement is within the Geauga County, Ohio area, comprising 156 church districts and 20,440 persons in 2023, having sprawled into three adjacent counties. The fifth largest Amish settlement, founded in 1850 in the Adams County, Indiana area, had nearly 10,000 fewer residents than the fourth largest settlement and less than a half of its church districts. In contrast with these large Amish settlements established in the eighteenth and nineteenth centuries, most Amish settlements are both considerably younger and much smaller. For example, the oldest Amish settlement in New York was established in 1949 and its second oldest settlement dates from 1974, yet that state had 58 settlements in 2023, the fifth largest number of any state. Missouri has 62 settlements, with its oldest being established in 1947. Minnesota's oldest Amish settlement dates from 1972 and Colorado's from 2002 (Young Center, 2023b). Of the Amish settlements in existence today—and many Amish settlements were relatively short-lived as they failed to attract sufficient newcomers to survive (Luthy, 1986, 2000)—60.8% were established since the beginning of 2000. About 87% are less than 50 years old (Young Center, 2023b).

The majority of Amish settlements, typically named for the nearest community with a post office, are small. While four Amish settlements exceed 100 church districts or congregations, most settlements are far smaller. Half (49.6%) of all Amish settlements in North America include only one church district. Another 124 settlements had only two church districts (Young Center, 2023b). Thus, 80.2% of all Amish settlements have only one or two church districts. Yet most Amish persons live in larger settlements. The majority of the small settlements are relatively young—no more than one or two decades old.

The establishment and growth of new settlements can be attributed to a combination of push and pull factors. Population growth among the Amish, who are doubling in population every 20.5 years (Donnermeyer, 2015), is a result of "large families [that] are an expression both of religious convictions and of a people whose economy is based on agriculture and other manual trades where the labor of children is valued" (Donnermeyer et al., 2013, p. 74). This growth has caused a dramatic increase in the price of farmland in the vicinity of the large Amish settlements, in particular, within the Lancaster County vicinity, which is also experiencing pressures resulting from urban—non-Amish—growth. Similar concerns are voiced in southwest Ontario, the home of Canada's oldest Amish settlement, established in 1825. Arsenault (2021, p. 26) notes, "There is strong correlational evidence that the rise of farmland prices is the key driver of Amish migrations, first out of Southwestern Ontario, then out of Central Ontario." The Amish who founded the Stanberry, Missouri Amish settlement explained that they sought to leave Michigan as they were "a bit unsettled mostly due to heavy population, high priced land, building codes, etc." (Miller & Miller, 2005, p. 249).

Locations that have farmland available for sale at reasonably low prices attract the interest of Amish families seeking to relocate (Anderson & Kenda, 2015; Kraybill et al., 2013). Locations with declining rural populations often have many farms on the market, sufficient to interest at least several Amish buyers. Other attributes are also desirable, including enough distance from a growing metropolitan area to minimize the prospect that suburban growth will escalate land prices. At the same time, being reasonably near urban markets in which to sell their produce and attract customers to Amish shops in their settlements is advantageous. In addition, access to nearby passenger rail service enhances the ability of Amish settlers to visit with kin and friends in the long-established Amish regions.

As Anderson and Kenda (2015, p. 486) explain, "To select a migration destination, Amish scouting parties explore several promising areas. Once group consensus emerges, households make land purchases near one another." For example, the Amish from Indiana who established the Bowling Green, Missouri settlement sought "several farms fairly close together so as to be able to drive together with horse and buggy" (Miller & Miller, 2005, p. 18). An Amish scribe explains that the Fremont, Michigan Amish site was "chosen for its farming qualities and within reasonable distance of several other settlements we hoped to fellowship with" (Anonymous., 2006, p. 127). Another Amish writer noted that the site of the Cashton, Wisconsin Amish settlement, the state's largest, was viewed favorably as the site selection delegation sought "a rural area with considerable land available and not near large cities… [with] good bus and train service within reasonable distance" (Hershberger in Miller, 2002, p. 68). The delegation of Amish from Geauga County, Ohio and Buchanan County, Iowa found that "many vacant farms [were] available" and that the "farms were very cheap in 1965-66" (Hershberger in Miller, 2002, p. 68). By March 1966 ten Amish families had purchased farms in a settlement that had grown to include 400 families in 2022 (Stoltzfus, 2022). Today, not only does the Cashton Amish settlement have Amtrak passenger service, but so do the Kingston-Dalton and Wilton-Tomah settlements, Wisconsin's second and fifth largest settlements.

The Platteville-Darlington Amish settlement within southwest Wisconsin is spread across much of Lafayette County, overlapping into Grant and Iowa Counties. This Amish community, established in 1999 by Amish farmers from Lancaster County and the "upper valleys of Pennsylvania," including Sugar Valley (Stoltzfus, 2022), provided ample relatively inexpensive lands for those who might be selling very expensive farmland in their former homeland. Indeed, the population of Lafayette County is now less than what it was in 1860, and today the Amish comprise about 7% of the county's population, with Amish landholdings far greater than within other Wisconsin settlements. For example, the average Amish household within the Platteville-Darlington settlement owned 82.4 acres in 2022, roughly twice the statewide average in Wisconsin. Ten Amish farmers in this settlement own over 200 acres, with two of them owning more than 600 acres (Cross, 2022, 2025). Land that is far less than a third the cost of active farmland in Lancaster County is a clear pull factor.

Real estate agents often assist Amish delegations seeking information regarding farmland availability. Indeed, in the Amish narrative explaining their selection of the site of the Bowling Green, Missouri Amish settlement, a scribe noted, "Jacob [Miller] wrote a letter to the postmaster at Bowling Green asking for the name of a reliable real estate agent" (Miller & Miller, 2005, p. 18). Following correspondence with that individual, an Amish delegation chartered a bus to tour the area and explore its possibilities. In some cases, local officials actively promote their areas to the Amish. For example, one Canadian "province, Prince Edward Island, has arguably done more than any other jurisdiction in North America to attract and retain Amish migrants" (Arsenault, 2021, p. 23). Farmland in Prince Edward Island, whose oldest Amish settlement was established in 2016, is about a tenth of the cost of land in southwest Ontario, the location of Canada's oldest and largest Amish settlements. The Amish who visited the province were courted by government officials. The province's school act was amended to accommodate Amish schooling practices. The provincial government developed a means to provide Amish with health care, without use of normal insurance, "given their willingness to take on agriculture, the province's number one economic sector, they promised to revitalize rural eastern PEI" (Arsenault, 2021, p. 35) and to promote tourism.

Lands owned and occupied by the Amish are widely scattered across their settlements, typically interspersed among properties owned by others. Using the Cashton, Wisconsin Amish settlement as an example, its households owned 19,554 acres scattered across 110 survey sections in 2021, thus on average 177.8 acres per square mile section. In fewer than half (46.4%) of these sections did the Amish occupy more than a quarter of the acreage of the section, and in only 15.5% did they own at least half of the section's acreage. Although in two sections the Amish owned all of the land except for the highway right-of-ways, elsewhere in Wisconsin in only two settlements are there single sections where Amish farmers own at least 550 acres of the 640-acre sections (Cross, 2022, 2025). Amish households comprise nearly half of the residents of the Town of Kingston (Cross, 2024), owning 3333 acres within the town's 31 sections (Fig. 7).

In the long-established and much larger Amish settlements of Lancaster County, Pennsylvania; Holmes County, Ohio; and Elkhart and LaGrange Counties, Indiana one can find sections where the Amish dominate land ownership. Amish individuals owned nearly 110,000 acres of farmland in Lancaster County (Rutter, 2004) two decades ago, yet cost of land and lack of available lands on the market necessitate expansion away from the core, creating a dispersed pattern of Amish landownership similar to what is seen in most Amish settlements. Even within Wisconsin's oldest Amish settlement, Medford, established in 1920 with an estimated population of

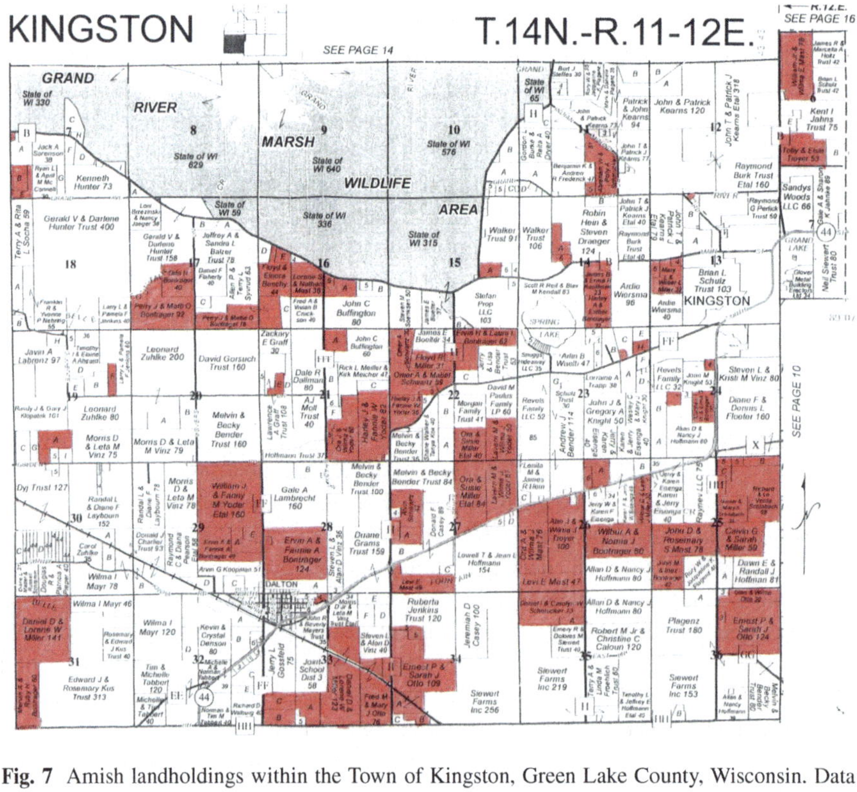

Fig. 7 Amish landholdings within the Town of Kingston, Green Lake County, Wisconsin. Data from Rockford Map Publishers (2020) and Yoder (2018). Copyright map from *Land Atlas & Plat Book, Green Lake County, Wisconsin, 2020*, page 8. Reproduced with Permission of Rockford Map Publishers, Inc., Rockford, Illinois

340 (Young Center, 2023c), Amish owned lands are spread across 24 survey sections. Yet in only two sections do Amish landholdings exceed 320 acres, or half of a section, while Amish landownership exceeds one-quarter of the land area in only 29.2% of the sections (Cross, 2025). Newer Amish settlements, whether the Westfield settlement in Wisconsin or the Ortiz settlement sprawling across the border between Colorado's Conejos County and New Mexico's Rio Arriba County, include fewer than ten Amish households extending across an area spanning six to seven miles (Stoltzfus, 2022).

The populations in which the Amish are interspersed vary widely, strongly influenced by the prevailing ancestry of the host counties. In no county do the Amish comprise the majority of the population. While the large and long-established Amish settlements in Pennsylvania are interspersed among many other Anabaptists, particularly Mennonites and Brethrens, as well as other German ethnics, the majority of Amish settlements are surrounded by individuals of other ethnic and religious backgrounds. For example, Wisconsin's largest Amish settlement was established on lands that had been long-occupied by Norwegian Lutherans, whose presence remains prominent within the area. That state's second largest Amish settlement occupies an area that includes many German Lutherans, along with smaller numbers of German Catholics and Welsh and Scottish ethnic Protestants. Old Order Mennonites are found in the vicinity of about a tenth of Wisconsin's Amish settlements (Cross, 2023). Amish settlements in Kentucky and Tennessee are located in areas with strong Scots-Irish Protestant ancestry, while Amish communities in Missouri are intermixed with both Scots-Irish and German ethnics. The Amish scribe describing the Ortiz, Colorado settlement noted the prominence of their Hispano neighbors and that the settlement's Amish school occupied "an old adobe house that we repaired" (Stoltzfus, 2022, p. 6).

Separate Society with Constrained Economic Collaboration

Religious segregation; education within their own parochial schools; conversations in Pennsylvania German rather than English; avoidance of radio, television, and social media; lack of intermarriage outside the group (with constraints even between different Amish affiliations); social interactions focused upon family, extended kin, and church members; and limitations on economic interaction with the "English" all interact to sustain the separate yet parallel existence of the Amish amid their modern more-worldly neighbors. Their separate reality has long been noted, and it is even more conspicuous as the Amish continue embracing horse-drawn buggies and plows, well over a century after their neighbors abandoned them. Eighty years ago, in his seminal study of the Lancaster County Amish, Walter Kollmorgen (1942, p. 21) noted, "The principle of separation from the world is still vigorously maintained by the Old Order Amish." This remains true. As Luloff et al. (2002, p. 157) note, "To the Amish, all outsiders are unbelievers, and intimate relationships are prohibited." Recently, it was reported that "Conservative Amish practice strict discipline maintaining greater separation from the non-Amish world" (Colyer et al., 2022, p. 34).

The local public high school, along with its associated interscholastic athletic activities, promotes cohesion within many rural communities and further links them to other communities within their region and state. Yet, the Amish lack these connections with their non-Amish neighbors by attending their own separate parochial schools, many of which accommodate all eight grades within one or two rooms. Likewise, their non-Amish neighbors worship elsewhere (or not at all) and often support athletic teams—whether public school, college, or professional—that are outside the world of the Amish, whose avoidance of radio and television leaves these sports activities unappreciated. Furthermore, the Amish are not members of local social clubs and professional associations, such as Kiwanis and Lions Clubs, that engage many of the "English". The whole range of entertainment that surrounds the "English", permeating their everyday interests and conversation, is absent from the lives of the Amish.

Employment in farming occupations also separates the Amish from their non-Amish neighbors—even those who are farmers. While many non-Amish farmers—particularly those with larger operations—employ a variety of farmhands, many of whom are immigrants or foreign nationals, Amish farms are much smaller, relying upon their many children or kin to provide the necessary labor. The equipment utilized on the farms of both groups differs significantly in type and size (Cross, 2014, 2016). Many non-Amish rural residents earn most of their income from off-farm jobs—often commuting long distances daily to work, even though they may have fields devoted to raising corn or soybeans. Yet, it has been argued, based upon longitudinal study in the Lancaster County area, that the Amish "interact frequently with the English and are increasingly linked to the larger society. Economic ties, in particular, have become more numerous" (Bridger et al., 2001, pp. 75–76).

In several of the largest Amish communities, while many Amish households live on small farms, fewer than one-in-ten are full-time farmers (Hurst & McConnell, 2010). In these places, particularly the Elkhart-LaGrange settlement of northwest Indiana, better than half of the male heads of households are employed in factory jobs. Amish working in factory jobs, often interacting with "English" workers, have a different take on their occupation compared with the "English." As Thomas Meyers (2022, p. 72) explains,

> Amish men work in factories to be able to come home and live comfortably in their own communities. 'Factory worker' is not their primary identity. For many people in the dominant culture, their work is the driving force in how they understand themselves and their relationship to those around them. For an Amish person, the answer to the question Who am I? is not related to the work that they do on a daily basis. It is much more related to nonwork hours, where one is primarily in an Amish context, relating primarily to Amish people and where Amish values are reinforced.

Nevertheless, research within the Holmes County Amish settlement of Ohio has shown that the engagement of Amish within non-farm business enterprises has created a less egalitarian Amish community (Moledina et al., 2014). Other Amish families have relocated to more rural locations, often in other states, when they wanted to remain in farming (Luloff et al., 2002).

Economic interaction between the Amish and their surrounding "English" community exists, yet it is constrained, particularly within the most conservative Amish settlements. The Amish need a market for their goods, produce, and labor. In many Amish communities one sees a variety of interactions to facilitate this marketing (Fig. 8), including the inviting of outsiders to patronize roadside produce stands, bakeries in a shop on their farmstead, woodworking shops that sell their furniture directly to non-Amish customers, and bulk grocery stores that cater to more outsiders than nearby Amish customers. Teams of Amish carpenters may take construction jobs in nearby communities. Nevertheless, none of these economic activities occurs on Sunday, and Amish shops—or signs on the farmstead—clearly state "No Sunday sales." Even the milk cans from an Amish dairy farm cannot be collected on a Sunday. Other Amish communities avoid interactions with the public at their farmsteads, selling their produce and products through intermediaries, such as a cheese factory that buys their milk, a produce auction house that sells their fruits and vegetables to wholesale buyers, or "English" stores that sell their furniture.

The Amish often must rely upon trusted outsiders to provide transportation, such as hauling milk to a cheese factory or transporting family members to a big box store for shopping. As mentioned earlier, the Amish often utilize passenger railroads to travel to distant settlements. Local "English" provide informal taxi service. This

Fig. 8 Sign advertising hand woven rugs and brown eggs for sale at an Amish farmstead visible in the valley. Note statement: No Sunday sales. Photo taken in the Boscobel Amish settlement within Crawford County, Wisconsin

author saw a local resident pulling the Amish church wagon behind his pickup truck within a church district where Amish farmsteads were spread across a considerable distance. Groups of Amish will sometimes charter a bus to visit relatives, attend a funeral or wedding, search for farmlands that can be purchased, or vacation. Yet, the Amish are travelling together, not mixing with outsiders except for the driver.

The Covid-19 pandemic provided an excellent example of how the response of the Amish, a classic example of a closed religious community, differed from that of the population at large. In noncompliance with Center for Disease Control (CDC) guidelines that religious services be conducted remotely, all of the Amish communities for which church information was reported by Amish scribes in Pennsylvania and Ohio settlements in *The Budget*, a prominent weekly publication, met in person at least once during the first 2 months of the pandemic restrictions (Concoran et al., 2022). However, 21% reported alterations, such as social distancing of families from other families, not serving the normal meal following worship, relocating a Sunday service to another farmstead given illness of the scheduled host, and avoiding the shaking of hands. Even for Amish congregations that were aware of the CDC guidelines, "restrictions on gathering [were] particularly burdensome…[given they] cannot shift to online services due to their sacramental limits on technology use" (Concoran et al., 2022, p. 4274). The pandemic also illustrated differences between the Amish and their neighbors in obtaining and paying for medical care and dealing with the financial costs of the shutdown.

Amish workers neither pay Social Security taxes (unless employed by a non-Amish employer) nor receive Social Security payments for either retirement or disability. Likewise, they do not have health insurance, whether provided via Medicare, Medicaid, the state, or private companies. Indeed, the Amish do not acquire any insurance, whether for health, life, fire, or liability, although Amish mutual aid associations often help Amish individuals and families pay medical bills and recover from disasters. Federal pandemic relief programs created a dilemma for the Amish. As Joseph Harasta (2021, p. 48) reported,

> Some Amish businesses never recovered from the lost revenue, mostly because they were unable or unwilling to accept direct government benefits, believing that if they took such payments they might jeopardize their exemption from Social Security and Medicare. Similar concerns produced some community discussion of how to return federal stimulus checks.

Amish reticence to receive government assistance, including use of food stamps, does not exclude their willingness to assist their "English" neighbors. Amish men serve as members of volunteer fire departments in some communities. Following major natural disasters, groups of Amish volunteers have travelled to disaster sites to lend their labor in both clean-up and reconstruction.

The majority of the Amish avoid involvement in both politics and litigation. Politically the Amish are largely non-voting, although both the Bush and Trump campaigns made efforts to reach out to them. While fewer than 1% of the Amish in Wisconsin's second largest settlement voted in the 2020 Presidential election, over 17% of Amish adults voted in Lancaster County, Pennsylvania (Kopko, 2023,

p. 119). Because most Amish are not registered voters and they are unwilling to swear oaths, Amish do not serve on juries. The Amish try to avoid litigation, and rarely take legal action against individuals who victimize them. Indeed, often "the Amish did not wish to pursue prosecution, ... were 'forgiving,' and wished to leave justice to God" (Byers, 2008, p. 239). Conflict between members of their communities is typically left to the bishop to resolve, which in some cases of domestic sexual abuse have led to controversy when local authorities have become involved. Conflict with local authorities regarding land use regulations, horse pollution, and environmental issues has occurred in some places, and desires for conflict avoidance does influence Amish decisions regarding locations of where to settle or vacate. As conscientious objectors, the Amish do not serve in the military.

Conclusion

Amish settlements display a constant interplay of separation and accommodation of distinct populations—the Amish and their "English" neighbors, whose collective display of modernity overrides any ethnic or racial differences that they have. Where accommodation becomes too onerous to the Amish, a settlement may lose members who seek other locations which are more tolerant. Academics have long speculated about the prospects for long-term survival of the Amish as a separate society existing within the midst of modernity (Foster, 1984; Kraybill & Olshan, 1994). Yet not only have they survived, but their numbers and spatial distribution have grown and continue to grow. Indeed, in under a decade (2014–2023) the number of Amish settlements in the United States increased from 465 to 617, while those in Canada grew from 15 to 20 (Young Center, 2023d).

References

Anderson, C., & Kenda, L. (2015). What kinds of places attract and sustain Amish populations? *Rural Sociology, 80*(4), 483–511.

Anonymous. (2006). *Michigan Amish directory 2006*. Abana Books Ltd.

Arsenault, G. (2021). Understanding Amish migrations to New Brunswick, Prince Edward Island, and Manitoba. *The Journal of Plain Anabaptist Communities, 1*(2), 22–41. https://plainanabaptistjournal.org/index.php/JPAC/article/view/7944/5924

Bridger, J. C., Luloff, A. E., Ploch, L. A., & Steele, J. (2001). A fifty-year overview of persistence and change in an old order Amish community. *Journal of the Community Development Society, 32*(1), 65–87.

Byers, B. D. (2008). Amish victimization and off Amish victimization and offending: A rural subcultural subculture's experiences and responses to crime and justice. *Southern Rural Sociology, 23*(2), 226–251.

Colyer, C. J., Stein, R. E., Corcoran, K. E., & Mackay, A. (2022). Amish population pyramids: Demographic patterns across affiliations in the Holmes County, Ohio, settlement. *The Journal of Plain Anabaptist Communities, 3*(1), 30–53. https://plainanabaptistjournal.org/index.php/JPAC/issue/view/322

Concoran, K. E., Stein, R. E., Colyer, C. J., Guthrie, S. K., & Mackay, A. M. (2022). Rituals of contagion in closed religious communities: A case study of Amish and Mennonite communities in the USA during the beginning of the COVID-19 pandemic. *Journal of Religion and Health, 61*, 4260–4281.

Cross, J. A. (2014). Continuity and change: Amish dairy farming in Wisconsin over the past decade. *The Geographical Review, 104*(1), 52–70.

Cross, J. A. (2016). Dairying landscapes of the Amish in Wisconsin. *Material Culture, 48*(2), 16–31.

Cross, J. A. (2018). Occupation patterns of Amish settlements in Wisconsin. *Journal of Amish and Plain Anabaptist Studies, 6*(2), 192–212.

Cross, J. A. (2021). Dairy woes in Wisconsin: What about the Amish? *The Journal of Plain Anabaptist Communities, 1*(2), 1–21.

Cross, J. A. (2022). Landownership within Amish settlements in Wisconsin: An exploratory study. *The Journal of Plain Anabaptist Communities, 3*(1), 13–29.

Cross, J. A. (2023). Old order dairy farmers in Wisconsin. *The Geographical Review, 113*(1), 72–90.

Cross, J. A. (2024). Sameness within the Amish settlements in Wisconsin. In S. D. Brunn (Ed.), *Geography of time, place, movement, and networks* (Mapping heritage journeys and sameness) (Vol. 2, pp. 47–65). Springer.

Cross, J. A. (2025). Amish land ownership within Wisconsin. *The Geographical Review, 115*(3), 370–392.

Crowley, W. K. (1978). Old order Amish settlement: Diffusion and growth. *Annals of the Association of American Geographers, 68*(2), 249–264.

Donnermeyer, J. F. (2015). Doubling time and population increase of the Amish. *Journal of Amish and Plain Anabaptist Studies, 3*(1), 94–109.

Donnermeyer, J. F. (2023). A demographic profile of the greater Lancaster County, Pennsylvania, Amish. *The Journal of Plain Anabaptist Communities, 3*(2), 1–34. https://plainanabaptistjournal.org/index.php/JPAC/article/view/9154/7875

Donnermeyer, J. F., Anderson, C., & Cooksey, E. C. (2013). *Journal of Amish and Plain Anabaptist Studies, 1*(1), 72–109.

Ericksen, E. P., Ericksen, J. A., & Hostetler, J. A. (1980). The cultivation of the soil as a moral directive: Population growth, family ties, and the maintenance of community among the old order Amish. *Rural Sociology, 45*(1), 49–68.

Foster, T. W. (1984). Separation and survival in Amish society. *Sociological Focus, 17*(1), 1–15.

Granville, K., & Gilbertson, A. (2017, September 15). In Amish Country, the future is calling. *New York Times*. https://www.nytimes.com/2017/09/15/business/amish-technology.html#:~:text=%E2%80%9CWe're%20not%20supposed%20to,how%20do%20you%20balance%20that%3F%E2%80%9D

Harasta, J. (2021). Pandemic among the Amish: A look at the effects of COVID-19 on Amish communities in Lancaster, Pennsylvania, in the first four months. *The Journal of Plain Anabaptist Communities, 2*(1), 42–61.

Hostetler, J. A. (1987). A new look at the old order. *The Rural Sociologist, 7*(4), 278–292.

Hurst, C. E., & McConnell, D. L. (2010). *An Amish paradox: Diversity and change in the world's largest Amish community*. Johns Hopkins University Press.

Kollmorgen, W. M. (1942). *Culture of a contemporary rural community: The old order Amish of Lancaster County, Pennsylvania. Rural life studies 4*. U.S. Department of Agriculture, Bureau of Agricultural Economics.

Kopko, K. C. (2023). Plain politics: Assessing old order Amish voter participation in the 2004, 2016, and 2020 presidential elections. *The Journal of Plain Anabaptist Communities, 3*(2), 105–126. https://plainanabaptistjournal.org/index.php/JPAC/article/view/9501/7880

Kraybill, D. B. (2010). *Concise encyclopedia of Amish, Brethren, Hutterites, and Mennonites*. The Johns Hopkins University Press.

Kraybill, D. B., Johnson-Weiner, K. M., & Nolt, S. M. (2013). *The Amish*. The Johns Hopkins University Press.
Kraybill, D. B., & Olshan, M. A. (Eds.). (1994). *The Amish struggle with modernity*. University Press of New England.
Land Atlas & Plat Book, Green Lake County, Wisconsin, 2020. Rockford Map Publishers.
Luloff, A. E., J. C. Bridger, & L. A. Ploch. (2002). The old order Amish community 50 years later. In A. E. Luloff & R.S. Krannich (Eds.), *Persistence and change in rural communities* (pp. 143–170).
Luthy, D. (1986). *The Amish in America: Settlements that failed, 1840-1960*. Pathway Publishers.
Luthy, D. (2000). *Why some Amish communities fail: Extinct settlements, 1961-1999*. Pathway Publishers.
Meyers, T. J. (2022). A demographic profile of the Elkhart-LaGrange old order Amish settlement. *The Journal of Plain Anabaptist Communities, 3*(1), 65–82. https://plainanabaptistjournal.org/index.php/JPAC/article/view/9157/7758
Miller, D. (2002). *Wisconsin, Minnesota, & Montana Amish directory 2002*. Abana Books.
Miller, D., & Miller, M. (2005). *Missouri Amish directory 2005*. Abana Books.
Moledina, A., McConnell, D., Sugars, S., & Conner, B. (2014). Amish economic transformations: New forms of income and wealth distribution in traditionally flat community. *Journal of Amish and Plain Anabaptist Studies, 2*(1), 1–22.
Noble, A. G. (1986). Landscape of piety/landscape of profit: The Amish-Mennonite and derived landscapes of northeastern Ohio. *East Lakes Geographer, 21*, 34–48.
Nolt, S. M. (2015). *A history of the Amish* (3rd ed.). Good Books.
Petrovich, C. (2017). More than forty Amish affiliations? Charting the fault lines. *Journal of Amish and Plain Anabaptist Studies, 5*(1), 120–142.
Rockford Map Publishers. (2020). *Land Atlas & Plat Book, Green Lake County, Wisconsin, 2020*. Rockford Map Publishers.
Rutter, J. (2004). *Amish culture remains deeply rooted in the land*. https://lancasteronline.com/news/amish-culture-remains-deeply-rooted-in-the-land/article_0dfd7822-bb93-537e-82bdeabac77d98fe.html
Stoltzfus, J. (2022). *Amish & Mennonite settlements of America: A guide to North America's horse and buggy groups and a directory of die Botschaft scribes*. Masthof Press.
Umble, D. Z. (1994). Amish on the line: The telephone debates. In D. B. Kraybill & M. A. Olsham (Eds.), *The Amish struggle with modernity* (pp. 96–111). University Press of New England.
Yoder, F. (2018). *Wisconsin Amish directory 2018*. Schlabach Printers.
Young Center. (2023a). *Twelve largest Amish settlements, 2023*. Young Center for Anabaptist and Pietist Studies, Elizabethtown College. http://groups.etown.edu/amishstudies/statistics/twelve-largest-settlements-2023/
Young Center. (2023b). *Amish population, 2023*. Young Center for Anabaptist and Pietist Studies, Elizabethtown College. http://groups.etown.edu/amishstudies/statistics/population-2023/.
Young Center. (2023c). *Amish population in the United States by state, county, and settlement, 2023*. https://groups.etown.edu/amishstudies/files/2023/11/Amish-Pop-2023_by-state-and-county_updated2.pdf
Young Center. (2023d). *Growth of Amish settlements and districts, 2014-2023*. https://groups.etown.edu/amishstudies/files/2023/11/Growth_2014-2023.pdf

Parallel Universes of the Two Koreas Through Comparative Mental Mapping

HaeRan Shin

Introduction

The notion of parallel universes has expanded beyond astrophysics and science fiction to describe coexisting but disjointed knowledge worlds. When communities occupy the same geographic or historical space yet remain largely isolated due to ideological, political or cultural barriers, their perceptions of each other can diverge to a striking degree. While parallel universes have been discussed in various contexts, Newton (2019) conceptualized the Cold War division space, including the two Koreas, as parallel universes. He saw the Cold War as a matter of parallelism and inter-penetration and distinct versions of the same or similar features including economic and political systems.

The Korean Peninsula, in which North Korea and South Korea are located, albeit in a divided state, is an articulated case of parallel universes. The two countries share a long history, linguistic roots, and geographical proximity, yet they have evolved into distinct societies disconnected by decades of political tension. The strain experienced by the two countries over the last 70 years in technical war status has prevented North Koreans from visiting, communicating or speaking with South Koreans. The lack of direct interaction or communication between the two reinforces their status as parallel worlds in which perceptions of the 'other' are shaped by second-hand narratives rather than lived experiences. Among the few exceptions to this separation include North Korean defectors who previously lived in North Korea and have since crossed the border.

H. Shin (✉)
Department of Geography, Institute for Korean Regional Studies, Institute for Gender Studies, Institute for Peace and Unification Studies, Seoul National University, Seoul, Republic of Korea
e-mail: haeranshin@snu.ac.kr

S. D. Brunn (ed.), *Transdisciplinary Research Across Knowledge Worlds*,
https://doi.org/10.1007/978-3-032-19491-6_4

This chapter delves into this phenomenon through the lens of mental mapping (Gieseking, 2013) by focusing on the perspectives of young North Korean defectors and South Koreans. A mental map is a subjective, private, unique image of the larger environment resulting from the mental filtering, coding of sensory data, and knowledge of the area (Gould & White, 2012; Graham, 1976). This chapter explores the concept of parallel universes on the Korean Peninsula through mental maps drawn by young North and South Koreans currently living in South Korea. By employing a comparative mental mapping method, this chapter builds on Kevin Lynch's seminal work, *The Image of the City* (1960), which explores how individuals perceive and organize urban environments. Extending his and subsequent studies' frameworks, this study examines not only the physical and spatial dimensions of mental maps but also their geo-emotional and geopolitical underpinnings. These mental maps also serve as a performative tool that elicits rich narratives and reflections from participants, thereby providing insights into how individuals construct and navigate their perceptions of two societies they either inhabit or imagine.

Drawing on participants' sketches of the two Koreas and follow-up interviews, the study investigates similarities and contrasts in their perceptions of both societies: for North Korean defectors, the country they have fled and the one they live in now; for South Koreans, the one they have never been allowed to visit and the one in which they currently live. While North Korean defectors have encountered both worlds firsthand, South Koreans' mental images of North Korea are knowledge worlds largely shaped by media and educational narratives rather than personal interaction.

By examining these mental maps, this research reveals how parallel knowledge worlds are constituted by ongoing geopolitical tensions, their experiences and imaginations and limited avenues for mutual understanding. Ultimately, this chapter argues that these mental maps are not just representations of space but are deeply embedded in the broader dynamics of identity, belonging and geopolitical tension. By exploring the parallels and divergences in the mental maps of young North Korean defectors and South Koreans, this study sheds light on the ways in which the Korean Peninsula operates as a divided yet interconnected parallel universe.

Parallel Worlds in Mental Mapping

Parallel universes or parallel societies have been discussed in the contexts of a digital society (Laskar, 2023), interventions in ghettos (Larsen & Delica, 2024; Secchi & Herath, 2021), homelessness (Speer, 2021), divided community (Čorkalo Biruški & Ajduković, 2012) and formal politics and informal modes of engagement (Pilkington & Pollock, 2015). These studies consistently find that such parallel worlds co-exist but rarely intersect or integrate; thus, they operate in isolation from each other. The notion of parallel universes contributed to our understanding of interconnected but vastly separated knowledge worlds and how different realities are constructed, sustained and transformed.

Geopolitical parallelism and national border crossing have recently received increased attention in discussions about parallel worlds. The emergence of far-right populism is based on the creation of parallel worlds based on social media (Skale, 2021). Media-populists and supporters have built a parallel world of their own perceptions including both anger and reproduced content (Oliveira & Castro, 2023). As Newton (2019) argued, the Cold War geopolitically divided geographies and maintained political and military tension. This division was particularly evident in the two Koreas. The division between North Korea and South Korea was enacted by the 1953 Korea Armistice Agreement, which respected the line separating the 1945 Soviet and US occupations. While interactions and encounters were not permitted, the two Koreas co-existed and influenced each other's development and strategies. International migrants, especially international students, were seen as living in a parallel world (Gomes, 2020; Spangler, 2023).

One critical feature of this parallelism is that people in one society are prohibited from visiting or directly interacting with those in the other. This enforced separation creates distinct imaginaries of "the other" that are constructed through media, education and second-hand accounts rather than through personal experience. For North Koreans, South Korea often exists as an abstract concept that is initially shaped by state propaganda and later reimagined through the secret consumption of South Korean media such as K-dramas. For South Koreans, North Korea is an unknowable land that is often framed through the lenses of fear, danger or pity as portrayed by government narratives and media. The inability of citizens from one Korea to visit those of the other cements this division rendering the societies truly parallel—adjacent but inaccessible, familiar yet foreign.

The terms "cognitive mapping" and "mental mapping" (Murray & Spencer, 1979) have been used interchangeably. These mental maps were useful in visualizing human-environment relations, psychological understanding and environmental behavior. Lynch (1960) introduced the concept of *imageability*, that is, the qualities of a physical environment that make it easily recognizable, memorable and navigable. Gould and White (2012, p. v) in their seminal 1974 book described the mental map as the "invisible landscapes that people carry in their heads." They developed cognitive representations of spatial environments that reflect individuals' perceptions, preferences, and knowledge of places, influencing spatial behavior and decision-making. The framework of the mental map provides a foundation for understanding how North Korean defectors and South Koreans create mental maps of North Korea and South Korea, particularly as these parallel worlds are deeply shaped by their physical, social and political boundaries.

Comparative mental mapping is a method that captures individuals' subjective perceptions of places by asking them to create visual representations or mental maps of two different societies, North Korea and South Korea. This approach is particularly insightful when applied to groups such as North Korean defectors and South Koreans whose spatial perceptions of each other's societies are shaped by their lack of access to the other's cultures and their own vastly different personal experiences. These maps are compared to identify overlaps, contrasts and the sociopolitical contexts underlying them. South Koreans' mental maps of North Korea are constructed through second-hand information, media and political discourse. North Koreans' mental maps of their own country, in contrast, reflect their images of North Korea rooted in nostalgia and personal memory while their images of South Korea are shaped by their adaptation to a new and unfamiliar environment. For South Koreans, the edge is an imagined boundary reinforced by political rhetoric and media. This boundary is absolute, that is, South Koreans cannot cross into North Korea, leaving their mental maps of the North abstract and speculative as they are shaped by external narratives rather than by lived experiences.

Mental maps are not static but dynamic, influenced by personal experiences, cultural contexts and social structures (Holmén & Götz, 2021). The act of drawing mental maps is also performative, revealing layers of meaning as participants articulate and reflect on their perceptions. The process North Korean defectors undergo when mapping their hometowns in North Korea often evokes emotional responses that transform abstract memories into vivid narratives. The mental mapping method not only explores spatial perceptions but also captures the psychological and emotional dimensions of living in or imagining parallel worlds (Thompson, 2020).

Research Methods: Mental Sketch Mapping and In-Depth Interviews

This research employs a comparative mental mapping method and interviews about the participants' mental maps. This method provides valuable insights into both the forms of spatial intelligence and the dynamics of human–environment relations in South Korea and North Korea in comparative terms. Through my previous work on North Korean defector entrepreneurs, I have established connections with both defectors and South Korean individuals who support them through various organizations. There are thirteen research subjects in total: seven North Korean defectors mainly engaged in businesses or employed in South Korea and six South Koreans professionally working on matters related to North Korean defectors. They belong to the age group of 20s to early 40s, and include five North Korean females, two North Korean males, three South Korean females, and three South Korean males.

I met them face-to-face individually, in locations where they felt comfortable, including their workplaces, coffee shops and my office. These South Koreans were selected because they potentially might have clearer mental images of North Korea due to their work than most South Koreans. South Koreans working in related organizations with North Korean defectors had a far more advanced understanding of North Korea compared to the average South Korean. More than 30,000 North Korean defectors live in South Korea including approximately 2000–3000 entrepreneurs among them. Table 1 summarizes the interviews and the participants' characteristics.

In one-on-one meetings, I asked the participants to draw one sketch about North Korea and one with South Korea as its subject. I said that they could draw a map, an

Table 1 Participants' characteristics

	Occupation	Age/sex	Hometown (North Korea)	Interview date	Interview place
North Korean participants					
1	Medical facility business	40s (F)	Pyunganbuk-do	9 Jan 2024	Coffee shop
2	Baby clothes business	30s (F)	Hwanghae-do	19 Jan 2024	Coffee shop
3	Journalist	30s (M)	Ryanggang-do	22 Jan 2024	Coffee shop
4	Matchmaking company	30s (F)	Ryanggang-do	7 Feb 2024	Her home
5	Coffee shop business	30s (F)	Ryanggang-do	8 Feb 2024	Coffee shop (her business)
6	Scientist	30s (M)	Pyongyang	28 Mar 2024	The author's office
7	Restaurant business	30s (F)	Hamgyongnam-do	4 July 2024	Restaurant (her business)
South Korean participants					
1	Counsellor (Hana Centre)	40s (F)		23 Jan 2024	Hana Centre coffee shop
2	North Korean alternative school	40s (F)		29 Feb 2024	Her office
3	Social worker	30s (M)		28 Mar 2024	The author's office
4	Hana Centre	30s (M)		3 July 2024	Hana Centre
5	Postgraduate student	20s (F)		8 Jan 2025	The author's office
6	Ministry of Reunification	40s (M)		10 Jan 2025	Office

abstract sketch, a landscape drawing or anything else they wanted to depict. South Koreans, as noted above are not allowed to visit North Korea, so their perceptions of the North were largely products of their imagination. Despite the small number of research subjects, one noticeable gender difference between those from North Korea and South Korea was observed: women provided detailed hometown landscapes, while men drew symbolic pictures such as black rainbows or Baekdu Mountain to represent North Korea.

One notable difference between the sketches of the different generation groups was the amount of information the North Koreans had about South Korea before their defection. While North Korea has a politically symbolic meaning and is mentioned frequently in South Korean media and education, South Korea was rarely mentioned in North Korea media. Many defectors, especially in their 40s and more that I observed in my previous studies, admitted to knowing almost nothing about South Korea before arriving. Their decisions to defect were often not based on specific knowledge of South Korea but instead influenced by routes that their brokers designed. However, young North Koreans, the subjects of this study, were influenced by the increasing popularity of K-dramas. These dramas, which were informally and secretly circulated among young North Koreans, portrayed affluent lifestyles and advanced living conditions that inspired their dreams of moving to South Korea.

These mental maps were performative in terms of developing the interview conversation. When asked to draw, the North Korean research subjects said, "Well, I'm not really good at drawing," to which I responded, "That's fine." However, as they started drawing, their narrative explanations of North Korea became significantly richer than they were without mental maps. Before the drawing activity, I asked about their hometowns, and their answers were quite brief. However, this drawing exercise became the starting point for them to provide more detailed descriptions of their hometowns. One participant who initially gave one-sentence answers about North Korea explained that she focused solely on her current job as a marriage counsellor. However, after completing her drawing, she began tearing up and confessed, "This is why I didn't want to talk about North Korea." She added that she missed her hometown deeply despite her more affluent life in South Korea. I selected and included some of these sketches and I analyze both their mental maps and narratives in the next sections.

North Korean Defectors' Mental Mapping of Experiencing Parallel Worlds

The drawings and narratives of the North Korean defectors demonstrated interesting contrasts and gaps between North Korea and South Korea and between their verbal explanations and mental maps. Houses were a central feature in their depictions of North Korea, as they were places in which people came together and established warm relationships. Their descriptions of these houses were quite detailed. By contrast, their drawings of South Korea featured coffee shops, restaurants, cars and airplanes, but did not include houses at all.

North Korean defectors have been allocated available council housing upon their arrival. In 2024, examining the regional distribution, Gyeonggi Province had the highest number with 11,080 residents, followed by Seoul with 6391, and Incheon with 2988. Jeju Island had the lowest number of residents at 358. They first live in public housing subsidized by the state, so their residence depends on the availability of public housing in various cities.

North Korean participants' verbal narratives emphasized the poverty of North Korea, but they were smiling, talking about how fun it was for people to get together for chatting and eating together. For example, North Korean Participant 1 said that North Korea is poor—there were few trees in the mountains, and women would gather by the river to wash clothes a common practice in earlier times before the widespread use of washing machines. At the same time, she elaborated how those women shared their daily life stories.

North Korean Participant 1's mental map of North Korea (Fig. 1) highlights the participant's memories of communal space. She described that after doing laundry, women in her hometown would go to one of their houses together. They and their children would enjoy chatting further. Although her sketch emphasizes bare mountains, modest houses with smoking chimneys, and a pastoral laundry setting, her narratives revealed cherished memories about her sense of community or familial closeness. For the interview subject, North Korea represents a place of warmth. This drawing reflects the participant's nostalgia for a place of quiet and modest living that contrasts sharply with the urban complexity she likely associates with South Korea.

North Korean Participant 1's mental map of South Korea (Fig. 2) portrays many cars, which symbolizes affluence and mobility, contrasting sharply with her depiction of North Korea. This depiction symbolizes modernity, mobility and the structured infrastructure of South Korea. She explains that there are many trees in the mountains and many cars in South Korea. As I looked at her drawing of people meeting in a coffee shop on her mental map, I asked why there were no houses in her sketch. She said:

> People in South Korea don't go to others' houses, do they? People always meet in a coffee shop. (North Korean Participant 1)

The cafe and no-house scene contrasts with the relational dynamics depicted in her North Korean map, which focused on familial and communal connections in domestic or natural settings. This reflects the participant's perception of a society in

Fig. 1 North Korean Participant 1's mental map of her hometown in North Korea: poverty but rich in human connecting

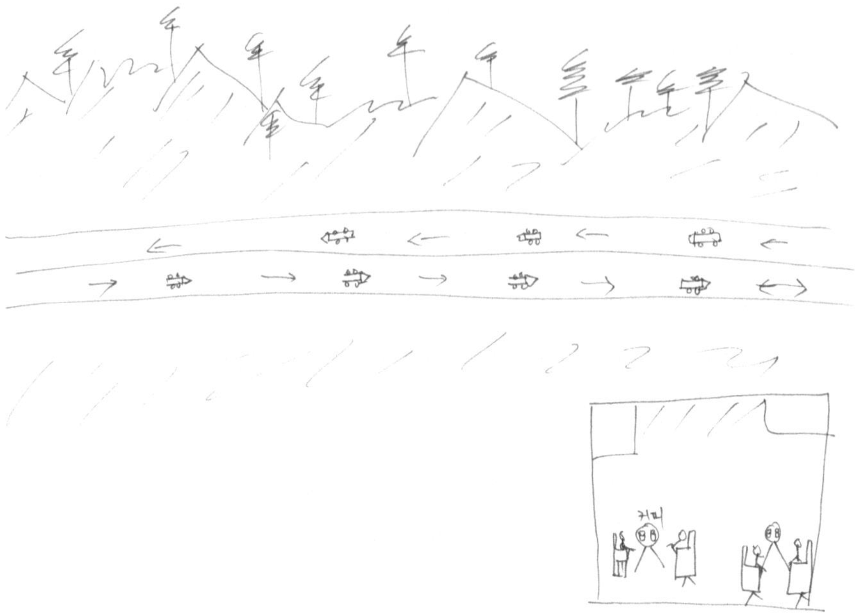

Fig. 2 North Korean Participant 1's mental map of South Korea: An affluent, mobile society with no houses

which relationships and interactions are more formalized and occur in public and commercialized environments.

North Korean Participant 2's mental maps and narratives were similar. She drew her hometown in detail (Fig. 3).

Figure 3 and her subsequent narratives capture both spatial and emotional significance. At the center of the map is her building and the next one, carefully detailed as multiple houses. She explained:

> Sorry. I have to draw these two apartment buildings because I knew EVERYONE in the building… The walls were thin, and after the head of the household (typically the man) left for work, if someone knocked on the wall, it was a signal to come over. Women and children would then gather in that house to eat and chat together. (North Korean Participant 2)

The map reveals a deeply communal culture in which relationships and shared spaces are integral to everyday life. In North Korean Participant 2's memory, gathering with friends and neighbors was central, and all the residents in her building knew each other very well. There was another house, labelled 'Friend's House'. She wrote, 'I came back here after spending a whole night with my neighbor friends.' The paths extending from the house to various locations reveal the spatial layout of her town and its accessibility. For instance, the *Jangmadang* market, a black market at that time that became legal in 2003, is an essential space for both economic activity and social interaction. It is reachable by bicycle within 15 min or on foot within an hour.

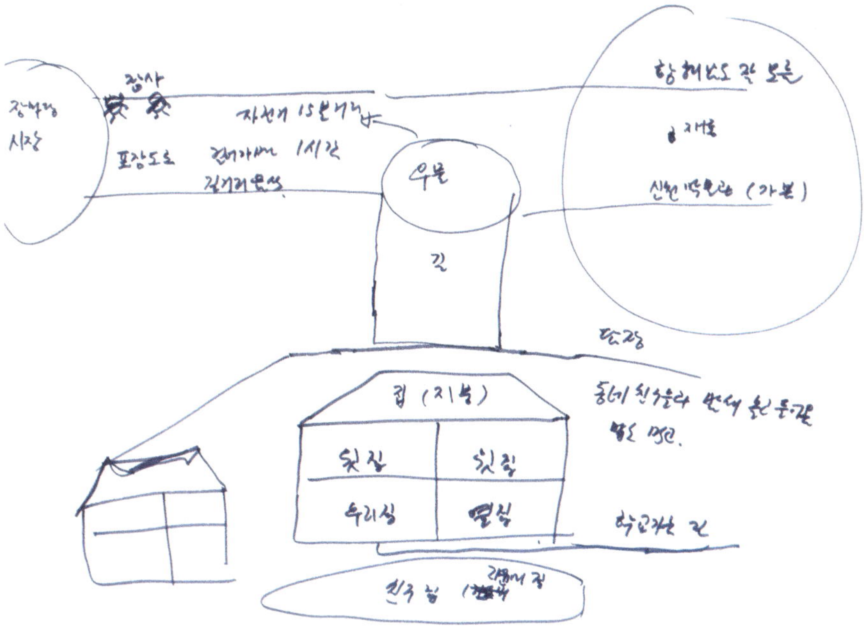

Fig. 3 North Korean Participant 2's mental map of her hometown in North Korea: neighbors, friends and *Jangmadang*

This memory contrasts sharply with her depiction of South Korea which highlights the differences in how space and relationships are experienced in these parallel worlds. Her mental map of South Korea (Fig. 4) reflects a depiction of an urban, structured and systematized society, capturing the participant's perspective of the country. This contrasts sharply with the personal, relational depictions of North Korea seen in the earlier map, offering insights into how the participant views South Korea as a space of modernity, mobility, competition and opportunities.

The airplane symbol suggests the modern, fast-paced and globalized nature of South Korea. As domestic and international mobilities are quite constrained, the number of transport facilities and the freedom of mobilities was one of the strongest impressions North Korean defectors had regarding South Korea. This participant's mental map of South Korea represents mobility, innovation and technological advancement, all of which are characteristics that align with South Korea's image as a leading global economy. North Korean Participant 2 wrote that there are many people in subways (grids), suggesting that someone has to be faster and work harder to succeed, thereby signaling the importance of surviving in competition.

On the right-hand side, the depiction of books or documents highlights education, knowledge and the structured acquisition of information. The text written close to this drawing suggests a society that provides opportunities, thus indicating a favorable system for early entrepreneurs. Unlike the warm, relational connections depicted in the North Korean map, this South Korean map appears more focused on

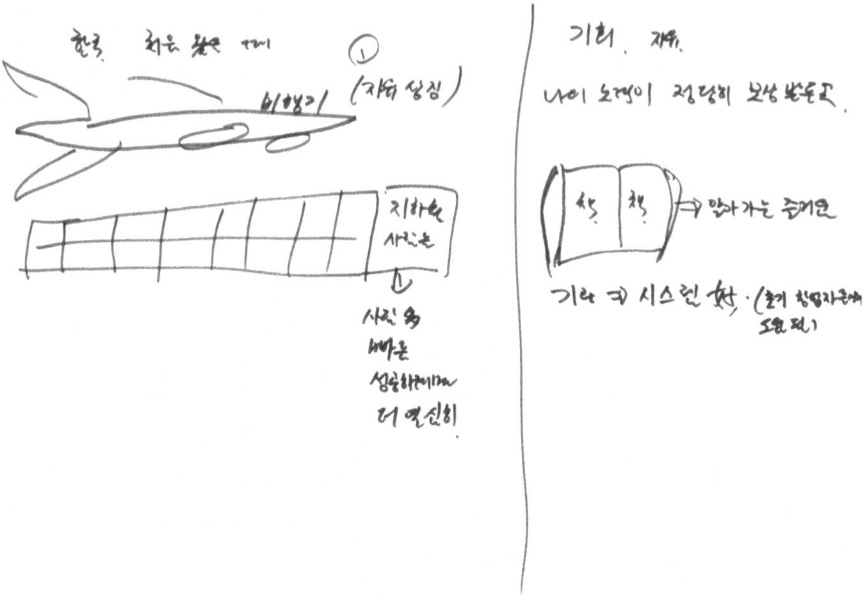

Fig. 4 North Korean Participant 2's mental map of South Korea: mobility, competition and opportunities

abstract or external markers of progress and modernity. This could reflect the participant's perception of South Korea as a place of urban complexity in which social interactions are less organic and more regulated than in North Korea. This highlights the tension between modernity and relationality in the participant's perception of the parallel words of the two Koreas. The divergence in these mental maps suggests that spaces play a crucial role in shaping identity and belonging.

Figure 5 represents the hometown of North Korean Participant 5 and provides a rich visualization of her memories and perceptions of the space. At the center of the map is a house, which is depicted as a modest structure with a door and windows. She explained that the structures surrounding and connected to the house are apartment buildings in which her family and friends lived. In the foreground, the Yalu River (*Amnokkang*) flows prominently across the map. This river is a significant geographical and symbolic feature that often serves as a boundary and lifeline for many communities in North Korea. It highlights the contrast between her rooted and relational experience in North Korea and the potentially more fragmented experiences she may have had in South Korea.

Her mental map of South Korea (Fig. 6) had no house again. The coffee shop, which was her business, was in the center. She wrote 'people,' but unlike her drawing of North Korea, which had many people, there is only one person in the South Korean map. This person travels with family in a car, symbolizing mobility and freedom. It offers a perspective on how North Korean Participant 5 perceives South Korea as a space of freedom, modernity and opportunities while also highlighting

Fig. 5 North Korean Participant 5's mental map of her hometown in North Korea: peoples' connections

Fig. 6 North Korean Participant 5's mental map of South Korea: coffee shop and mobility

its contrasts with their hometown in North Korea. The map also includes the labels 'people', 'work colleagues', and 'good people', reflecting the participant's focus on relationships and connections in South Korea. These descriptions suggest that despite its urban complexity, South Korea is seen as a place in which meaningful interpersonal connections can be formed, particularly in professional and social settings.

This distinction implies that parallel worlds not only separate people geographically but also reshape their social norms and interactions. The inability of defectors to connect with South Korean domestic culture reveals a form of estrangement as they struggle to bridge the gap between their previous communal practices and South Korea's urban individualism. In their personal lives outside of work, they interact little with South Koreans while maintaining many connections with other North Korean defectors, regularly visiting each other's homes. Many North Korean defectors encounter South Koreans primarily at work and through churches, where they receive both financial and religious support. In parallel worlds like North Korea and South Korea, spatial practices evolve independently due to differing political, economic and cultural trajectories. This divergence underscores how parallel worlds develop distinct spatial priorities based on their socioeconomic systems—North Korea's communal and resource-constrained environment versus South Korea's globalized and consumer-driven urban context.

Houses were central in North Koreans' drawings of their homeland. The drawings of these houses were followed by narratives regarding gatherings with friends and families reflecting the strong emphasis on familial and community ties in North Korean society. Homes in North Korea are depicted as spaces of intimacy, social gatherings and belonging. These drawings suggest that interviewed defectors associate North Korea with a deeply personal and relational sense of space.

By contrast, the absence of houses in their South Korean mental maps and the emphasis on transport and public spaces like coffee shops and restaurants demonstrate the economic advancement of South Korea, its freedom of mobility and formalized meetings among people. They also highlight a cultural shift, as early in the twentieth century, as such gatherings were as common in the South as they were in the North. In contemporary South Korea, social interactions often take place outside the home reflecting a more individualistic and urbanized lifestyle. This shift implies a transformation in how people connect with each other with less emphasis on private domestic spaces as sites of community.

For North Korean defectors, the absence of houses in their depictions of South Korea may also signal a sense of alienation or disconnection. This means that parallelism existed not only between South Korea and North Korea but also between the defector's former society and the South Korean society where the defector lives now. Homes are traditionally viewed as spaces of comfort and identity, but the defectors' inability to associate South Korean houses with similar functions suggests that they may feel excluded from or unfamiliar with the domestic culture of South Korea.

South Koreans' Mental Mapping of Imagining Parallel Worlds

South Koreans' mental maps were more abstract than North Koreans' ones, which reflects the lack of their experience with the latter society. They were often critical and cynical about both South Korea and North Korea. For instance, they mentioned people in South Korea as being controlled by competition and conscious of others' judgments to the point of faking their realities while North Korea was described as heavily influenced by government control and characterized by images of violence and unpredictability. Except for one South Korean who had visited North Korea, South Korean participants did not mention the differences of mobility as they took the freedom of mobility for granted.

As the South Korean participants worked closely with North Koreans, they were sympathetic towards North Korean defectors and aware of the harsh and negative realities many defectors face. Their mental maps of North Korea were either surprisingly positive or cynical—as much as their maps of South Korea were.

South Korean Participant 3, a social worker working with North Korean defectors, sketched North Korea (Fig. 7). He watches many YouTube videos involving a girl who talks about her luxurious apartment and Lotte in North Korea to promote the nation. He said:

> I remember the propaganda. But they only show a façade of luxury while behind the scenes, they're always with no freedom—this is what came to mind. (South Korean participant 3)

South Korean Participant 3's sketch (Fig. 8) and narratives about South Korea demonstrated another subject faking reality. He said South Koreans go out and turn the camera on for Instagram. He said:

Fig. 7 South Korean Participant 3's mental map of North Korea. He watched a reality faked by the state

Fig. 8 South Korean Participant 3's mental map of South Korea: people fake their lives by themselves

> People just live in small, run-down rooms at home, but when they eat, they take pictures of luxurious things and pretend to be affluent and happy on Instagram. (South Korean participant 3)

Eventually, there is much "sameness" in both North and South Korea: lives are faked even though the systems are different. The difference is that they are faked by the state in North Korea and faked by themselves in South Korea. The cynicism South Koreans express about their society, such as the emphasis on being overly conscious of others' judgments and faking realities, reflects a critical awareness of the social pressures embedded in South Korea's competitive, capitalist society. The observations in South Koreans mental maps underline the essence of parallel worlds: two coexisting but disconnected worlds that influence each other in profound yet indirect ways. The cynicism towards South Korean society reflects internal tensions within a modern, globalized society whereas the perception of North Korea as controlled and violent reflects an externalization of fear and otherness.

Another participant, South Korean Participant 2, a teacher in a North Korean alternative school, described the work as rewarding, viz., a feeling she struggled to find when teaching South Korean students raised in a capitalist society.

This participant explained and contrasted her two sketches (Figs. 9 and 10) as follows:

> North Korea is a barren land; it is cracked and dry. But the fruit is that the North Korean children are loving and caring. In South Korea, buildings are complex, crowded and somewhat desolate. The buildings seem to ignore each other, trapped in their own spaces and surrounded by junk. They're not really about love. (South Korean participant 2)

This participant was happy with her North Korean students, who lived and came directly from North Korea and not by living in a third country. She said that she was truly loved by and learned about love from those students. She described North

Fig. 9 South Korean Participant 2's mental mapping of North Korea: Good fruits -children- and barren land -the North Korean regime

Fig. 10 South Korean Participant 2's mental mapping of South Korea as being full of buildings that ignore each other

Korea as being poor and unproductive, despite North Koreans being loving, while South Korea is depicted as being advanced but lacking love. Other South Korean participants' mental maps similarly highlighted the social competition and hierarchy made as a result and opportunism of South Korea and the control and pretension of Kim Jong Un's family in North Korea.

Notably, a South Korean participant who had the opportunity to visit North Korea created a mental map between North Korean participants and South Korean ones. He emphasized freedom of mobility, much like the maps described by most North Korean participants. His mental map of North Korea (Fig. 11) highlights the limited number of cars and the geographical restrictions on car mobility in North Korea compared to the freedom to move anywhere in South Korea, a contrast also emphasized by the North Korean participants.

South Korean Participant 6's mental mapping of South Korea (Fig. 12) demonstrates the free mobility of the country's people, while the city seems complex and scattered.

Fig. 11 South Korean Participant 6's mental mapping of North Korea: A highly organized and mobility-constrained society

South Korean Participant 6 explained his impression of South Korea and his change in perspective as follows:

> When I was in Kaesong, North Korea and then crossed back into the South, I found it chaotic at first; however, I also noticed that people could move about freely in any direction they wished…. Despite its complexity, it's a place where people seem to thrive amid the hustle and bustle. (South Korean participant 6)

This statement meant that he did not notice the freedom of mobility before he saw North Korean cities. North Korea was a parallel that was highly different from the world he knew. He wrote:

> In contrast, when you look at North Korea, at first glance it appears very orderly. The city centers are incredibly well-organized—it gives off a 'new town' vibe. But, up close, they're unimpressive. Upon closer examination, you realize that everything is designed to keep people confined to their designated roles and spaces. (South Korean participant 6)

This statement contrasts with his description of South Korea. He said that job–housing proximity is positive in principle, but the North Korean version is a symbol of constrained mobility. His mental map was distinct from those of the other South Korean participants, which were quite critical of South Korea for being full of competition, hypocrisy and capitalistic values. His visit to North Korea impressed upon him the vast difference between the two Koreas in terms of mobility freedom.

Fig. 12 South Korean Participant 6's mental mapping of South Korea: Complex and highly mobile

Conclusion

The findings of this research show how North Korean defectors and South Koreans perceive the two parallel worlds through their direct experiences, social media and information that they obtain from their careers. Houses dominate the North Korean defectors' mental maps of North Korea, representing warmth, familial ties and community. Conversely, their depictions of South Korea focus on mobile lifestyles, affluence and competition. South Koreans' mental maps of North Korea, by contrast, are abstract and shaped by external narratives focusing on dictatorship and regulation, while their maps of South Korea reveal a mixture of pride and cynicism surrounding their society's competition-oriented atmosphere. For defectors, homes in North Korea represent a nostalgic connection to their roots and a sense of identity that they may struggle to replicate in South Korea. The absence of houses in their South Korean maps reflects a cultural and emotional gap as well as a possible sense of transience or displacement.

This study implies that the divisions created by the geopolitical forces of the Cold War have evolved into parallel worlds maintained by physical borders and different methods of how societies work. The concept of parallel worlds is particularly relevant in the context of the Korean peninsula. It highlights not only the physical separation of the two Koreas, but also the psychological and cultural divides that sustain their isolation. The prohibition on the interaction between North and South

Koreans, as well as their varying perceptions of the two Koreas, created a significant divergence in the mental maps of these two. The maps of North Korean defectors illustrate their experiences of the two societies and a deep sense of loss and longing for their inaccessible hometowns. South Koreans' mental maps pass judgment of the two societies' ideologies and economic systems and landscapes.

The research also suggests that the parallel worlds reflect not only the effects of physical and political separation and isolation but also connectedness and co-constitution. As the two Koreas leverage their division for their own economic and political strategies, they constitute each other as their respective main reference point. These contrasts highlight the role of enforced isolation in shaping perceptions and the powerful influence of constructed narratives in the absence of direct experience. This comparative mental mapping constitutes an important part of the reality of the parallel worlds and subsequently becomes reality itself.

Future research could explore several dimensions to further illuminate North Korean defectors' evolving spatial and social perceptions. Key areas include examining interactions between defectors and South Korean society through physical encounters and digital platforms, while considering generational differences in adaptation. Another avenue would be analyzing defectors' first photographic choices upon arrival in South Korea, offering insights into their initial impressions. Investigation of shared spaces—community organizations, religious institutions, and recreational sites—could provide perspectives on integration, particularly through specific Seoul-based groups facilitating such interactions. Museums featuring North Korean exhibits present another research opportunity, especially regarding material acquisition and narrative curation. Gender differences and marriages between North Korean women and South Korean men would be a potential topic. These topics will collectively deepen our understanding of the mental geographies shaping defectors' experiences and spatial identity reconstruction in the South.

Acknowledgement This work was supported by the Ministry of Education of the Republic of Korea and the National Research Foundation of Korea (2023S1A5A2A03083366) and the Institute for Peace and Unification Studies at Seoul National University.

References

Čorkalo Biruški, D., & Ajduković, D. (2012). Parallel worlds of divided community: Time does not make much difference. In O. Simić, Z. Volčič, & C. R. Philpot (Eds.), *Peace psychology in the Balkans* (*Peace psychology book series*) (pp. 177–198). Springer.

Gieseking, J. J. (2013). Where we go from here: The mental sketch mapping method and its analytic components. *Qualitative Inquiry, 19*(9), 712–724.

Gomes, C. (2020). Living in a parallel society: International students and their friendship circles. *Journal of International Students, 10*(1), xiii–xv.

Gould, P., & White, R. (2012). *Mental maps* (2nd ed.). Routledge.

Graham, E. (1976). What is a mental map? *Area, 8*(4), 259–262.

Holmén, J., & Götz, N. (Eds.). (2021). *Mental maps: Geographical and historical perspectives*. Routledge.

Larsen, T. S., & Delica, K. N. (2024). *Fragmenting cities: The state, territorial stigmatization and urban marginality*. Edward Elgar Publishing.
Laskar, M. H. (2023). Examining the emergence of digital society and the digital divide in India: A comparative evaluation between urban and rural areas. *Frontiers in Sociology, 8*, 1145221.
Lynch, K. (1960). *The image of the city*. MIT Press.
Murray, D., & Spencer, C. (1979). Individual differences in the drawing of cognitive maps: The effects of geographical mobility, strength of mental imagery and basic graphic ability. *Transactions of the Institute of British Geographers, 4*(3), 385–391.
Newton, S. (2019). Parallel worlds: Cold war division space. In M. Craven, S. Pahuja, G. Simpson, & A. Saunders (Eds.), *International law and the cold war* (pp. 117–136). Cambridge University Press.
Oliveira, A., & Castro, G. (2023). Media and democracy in Brazil: Advances and setbacks since the 2022 presidential elections. In I. F. Cunha, L. Guazina, A. Cabrera, & C. Martins (Eds.), *Media, populism and corruption* (pp. 127–144). ICNOVA.
Pilkington, H., & Pollock, G. (2015). 'Politics are bollocks': Youth, politics and activism in contemporary Europe. *The Sociological Review, 63*, 1–35.
Secchi, D., & Herath, G. B. (2021). Parallel society: Myth or reality? A question for policy makers. *Journal of Simulation, 15*(1–2), 148–158.
Skale, E. (2021). The rise of the new right: Psychoanalytic perspectives. In D. Morgan (Ed.), *A deeper cut: Further explorations of the unconscious in social and political life* (pp. 233–245). Phoenix Publishing.
Spangler, V. (2023). Home here and there: A spatial perspective on mobile experiences of 'home' among international students. *Social & Cultural Geography, 24*(8), 1440–1457.
Speer, J. (2021). Subalternity as displacement: Memoirs of homelessness and the struggle to be heard. *Environment and Planning D: Society and Space, 39*(4), 627–644.
Thompson, M. (2020). Mental mapping and multinational migrations: A geographical imaginations approach. *Geographical Research, 58*(4), 388–402.

The Parallel World of the Roma: From the "Straw Road" to the Ghettos and the Architecture of Exclusion

Krasimir Asenov and Irena Nikolova

Introduction

The postmodern world, shaped by globalization, digitalization, and cultural hybridization, promotes conditions for a global, homogeneous community. Yet some societies preserve their own cultural logic, social organization, and symbolic order, independent of the dominant models of modernity. These societies, which we may conventionally describe as "parallel" are not geographically isolated but culturally autonomous. A striking example is the Roma, a community existing for over 1500 years under systemic victimization—genocide, forced assimilation, discrimination, and social exclusion. Today, Roma ghettos remain "geographies of despair," layered with a syndrome of deprivation that constructs a multidimensional architecture of exclusion. Despite this, Roma communities sustain collective identity and survival through mobility, adaptability, narrative and oral-musical memory, and specific moral norms (*mahrime*) that uphold cultural autonomy.

This discussion examines Roma communities and ghettos as a "parallel civilization"—a social and cultural system that endures outside dominant modern institutions. Within this framework, the "Straw Road" symbolizes an existential structure bridging the territorial logic of modernity and the Roma cultural paradigm, rooted not in institutions or land but in movement, narrative, and living memory.

K. Asenov · I. Nikolova (✉)
University of Agribusiness and Rural Development (UARD), Plovdiv, Bulgaria

S. D. Brunn (ed.), *Transdisciplinary Research Across Knowledge Worlds*,
https://doi.org/10.1007/978-3-032-19491-6_5

Parallel Worlds and the Politics of Visibility

The parallel world of the Roma is not merely an isolated social fact but a complex system of embodied, spatial, and visual regimes. The Roma body, marked by historical traumas—slavery, the Holocaust, and institutional segregation—serves both as an ethnic marker and a mobile territory of exclusion. It carries a stigma inscribed through skin color, clothing, language, and movement, remaining hyper visible in public yet symbolically erased, "present through its absence" (Fanon, 2008, p. 94). This paradox is heightened by institutional humiliation: limited healthcare, controlled education, and segregated housing. Simultaneously, the Roma body is subjected to pervasive surveillance—from school registers to police checks, social workers to supermarket guards—rooted in historical patterns of racialization and state control, what Bauman (2004, p. 71) calls "management through surplus," regulating populations deemed superfluous or hard to integrate.

The Roma territory—*mahala*, ghetto, or informal settlement—functions as a spatial extension of biopolitics, a locus of concentrated exclusion—simultaneously a refuge from the racist outside world and a trap difficult to escape. As Henri Lefebvre (1991, p. 26) notes, space is never neutral but produced by social relations, power hierarchies, and economic logics. The *mahala* is thus a political space, shaped by historical marginalization and the spatial inscription of "otherness."

These parallel worlds exist not outside society but at their core, in a state of permanent exclusion. They are fragments of the social body that society refuses to acknowledge. In them, Mbembe's (2003, p. 11) "necropolitics" operates, governing marginalized bodies like Roma through violence—house demolitions, forced evictions, police raids—which is institutionalized rather than exceptional, constantly signaling "which bodies belong" and "which must be expelled."

The Roma parallel world also encompasses alternative economies, social structures, and norms, operating outside formal state logic. This "social smuggling of life" (Graeber, 2011) is not poverty but a resilient modality of being and resistance to the normative state. Yet it is often criminalized, with "Roma culture" blamed for poverty, unemployment, or "unsocialized" children. In this way emerges what Pierre Bourdieu (1999, p. 89) terms "symbolic violence"—the process through which the oppressed come to perceive their own exclusion as natural. Roma parallel reality persists not from refusal to integrate but because society enforces nonbelonging; their world exists as society's shadow—produced by it yet simultaneously denied by it.

For Roma communities, the concept of *axis mundi* (sacred space or sacred structure) takes on a stratified form. While macrosocieties rely on institutionalized, fixed symbols—squares, cathedrals, state buildings—Roma ghettos are often denied such access. Here, illegal home serves as a sacralized point of orientation, structuring world and identity. Demolition of these homes erases this local *axis mundi*, profoundly affecting the community's physical and symbolic foundations.

The loss of a fixed *axis mundi* in ghetto spaces does not erase the sacred center but transforms it. Amid mobility, insecurity, and periodic evictions, Roma communities reconstruct their cosmic center in mobile, immaterial forms. Caravans, central campfires, and collective practices like music and dance act as portable "axial lines," linking the community to symbolic order and sustaining cultural integrity. These mobile *axes mundi* require no territorial anchor yet perform the same essential function: organizing space, time, and social energy around a sacralized point, wherever the group may be.

In Roma culture, the turtle symbolizes stability and a portable home, serving as a symbolic *axis mundi*. Like the turtle, the Roma carry their "home" and cultural center wherever they go.

Music and dance are central to Roma culture, shaping identity and social life (Fig. 1). Beyond entertainment, they serve ritual functions, connecting the community to ancestors and traditions. Roma music creates a temporary sacred space—mobile *axis mundi*—that links present and past, sustaining communal spirit.

The juxtaposition of the *axis mundi* in macrosociety and Roma communities reveals two distinct models of sacred centralization. In the first, the center is institutionalized, architecturally anchored and symbolically reinforced through monumental structures that materialize the stability of the social order. In the second, the center is flexible, mobile, and often situated beyond the legal framework, yet it remains equally significant for the community's cultural and cosmological

Fig. 1 Engagement in the Roma neighborhood in Asenovgrad, Bulgaria. As a mobile ritual, Roma dance materializes the community's impermanent yet continuous center—a non-fixed *axis mundi* that emerges in each act of collective movement. (Source: Krasimir Asenov)

orientation. These parallel *axes mundi* not only reflect distinct historical and social conditions but also illustrate that the sacred center constitutes a universal human need, expressed through diverse spatial and symbolic forms.

Ethnographic Sketch: "The Straw Road"

The Straw Road is neither a metaphor nor a poetic image, but an anthropological reality that shapes the cultural autonomy of Roma communities in Europe as a parallel civilization, operating according to its own ontological, symbolic, and epistemological principles. This road is not a geographical line but a topological condensation of experienced deracination—the absence of home, land, institutions, and stable lineage. It is a route of survival, and simultaneously a path of resistance—not through weapons, but through cultural defiance, mobility, and refusal of fixation.

Within the framework of an ethnographic study in a Roma *mahala* in southeastern Europe, a conversation with an elderly Roma man, referred to here as Rashko, reveals this reality in its full emotional and historical depth. Sitting in his yard amid a chaotic yet functionally organized space of building materials covered with plastic sheets, he fed birds while recounting a story that belongs not solely to him, but to an entire cultural continuum. His narrative—spontaneous and fragmented, shifting between childhood memories and contemporary grievances—exemplifies a traumatic narrative (LaCapra, 2001), in which chronological logic is subordinated to affective and symbolic reasoning.

The story begins in the early twentieth century, during the Balkan Wars and the First World War—a period of mass migration, shifting borders, and the destruction of traditional communities. Rashko's family, part of a Roma *katun* (encampment), lived near a Greek village. One day, the father sent his 11-year-old son (Rashko's grandfather) to the village to ask for a few nails. When the boy returned, he found burned wagons, scattered belongings, and the disappearance of his relatives. After spending the night alone in an abandoned shack, he set off the next day on foot, directionless. On the road, the boy was taken in by another Roma *katun*, fleeing the pogroms. While traveling in a cart beside an elderly sick man, the child recounted what had happened. The old man comforted him: "If you follow the straw along the road, one day you will find your family." From that moment, the boy learned to search for and read the signs left by Roma along the roads—symbols marking food, danger, shelter, solidarity, and, above all, presence. After months of wandering, exhausted and ill, the child reached a suburban Roma *mahala* in southern Bulgaria, where he was given shelter. He remained there, married, and started a new family. "He never saw his relatives again," Rashko concluded, adding: "We Roma have lived for centuries along the Straw Road. Straw for the bed, for the fire, for the horses. After us, only straw remains."

This story is not merely an autobiographical memory; it is a cultural chronicle, where personal memory and collective trauma intertwine (Assmann, 2011; Ricoeur,

2004). It serves as a cultural document revealing several key themes: the transience of Roma mobility, the absence of institutionalized historical memory, and cultural forms of resistance expressed through orality, ritual, and the transmission of codes.

Straw, as both material and symbol, becomes a central bearer of meaning—not merely a means of survival, but a mnemonical marker (Connerton, 1989) that sustains epistemological resilience: "If you follow it, you will always find your own." Straw thus symbolizes the ontological flexibility of Roma identity—its capacity to adapt, avoid fixation, and leave no traces that can be archived or controlled.

The Straw Road is not simply a route from point A to point B—it is an archaeology of presence without possession, of memory without archive, and of resilience through withdrawal, as described by James Scott (2009) in *The Art of Not Being Governed*. It is not a linear trajectory but a structure of feeling (Williams, 1977, p. 131), rather a cultural vector through which the Roma community articulates its difference vis-à-vis hegemonic systems of memory, belonging, and territorialization.

The Straw Road diagnoses a parallel and alternative European presence (Van Baar, 2011), one that resists integration into national histories, as Roma history—mobile, oral, and post territorial—defies state archiving. Tracing migration from the Indus to the Rhine, it leaves no monuments but a persistent, elusive thread guiding Roma along Europe's cultural, political, and spatial margins. Roma mobility is not merely economic or forced; it is a form of cultural epistemology—knowing through experience and storytelling rather than mapping—aligning with Deleuze and Guattari's (1986) nomadic machine, which resists state inscription and territorial fixation.

The nomadic machine operates through rhythm, movement, repetition, and discontinuity. Likewise, the Roma cultural system functions via a rhythmic mesh (Ingold, 2011)—a network of recurring practices, performative markers, and localized strategies for survival and knowledge transmission—exemplifying territory without statehood (Lucassen et al., 1998): resilient yet decentralized, mobile yet coherent.

One of the cultural micro scenarios that encodes this parallel civilization is the following parable, orally transmitted in various versions across Roma communities in the Balkans.

> In a faraway kingdom, there once lived a king who was very lonely. One day, he decided and proclaimed that he would marry the most joyful girl in the world and make her his queen. The news spread to every corner of the earth. On the appointed day, at the set hour, hundreds of young women gathered in the great hall of the castle. All were covered with opaque veils, so the king could see neither their faces nor their figures. To the sound of music, he began to walk among them. Time passed, yet he could not bring himself to make a choice. At one moment, however, he noticed a slight, almost imperceptible movement—one of the maidens gently swaying, barely following the rhythm of the melody. She was not dancing; she was living the music. The king was certain—he had made his choice. When her veil was lifted, before him stood a smiling Roma girl. After their wedding, the palace was filled with songs, laughter, and life. Yet every night, when the king fell asleep, the young queen would quietly leave their chambers to sleep under the stars. One night, the king followed her and asked why she left the comfort of her bed. She replied: "If I cannot

> see the stars, I cannot fall asleep, my king." So, the king ordered a glass dome to be built over their bedroom, so she could gaze at the night sky even from her bed. And thus, they lived together—he in her world, and she in his.

This parable, more than folklore, encapsulates key aspects of Roma cultural logic: music acts not as background but as a mediator of recognition, revealing true identity. It underscores the difference between visually oriented cultures and those operating performatively and acoustically—through body, rhythm, and listening (Howes, 2003; Ong, 1982).

The veil symbolizes visual suppression—the inability of hegemonic culture to "see" the Other. Yet it is precisely through bodily resonance with rhythm that the subject is recognized. In this way, an alternative ontology of presence is articulated—not through fixed traits or institutional markers, but through sensitivity to vibration, rhythm, and life. This constitutes a wholly performative form of cultural assertion, one that resists the visually dominated regime of modernity (Goody, 1987). The desire to sleep under the stars is more than sentiment—it is a cultural paradigm, symbolizing openness to the cosmos and a refusal of the architecture of fixation. The sky here is the primordial home—a shared, unenclosed space that rejects regimes of ownership. The Roma cosmos unfolds as an ontology of the open (Nancy, 2000), in which existence is lived through limitlessness and rhythm rather than through territorial possession. The role of the king is symptomatic: institutional power does not impose coercion but enacts adaptation. Rather than assimilating the Other, it acknowledges the possibility of being oneself. The glass dome that is built functions as an architectural compromise—a synthesis of institutional recognition and ontological freedom. This is not utopia, but a model of coexistence that presupposes mutual sensitivity and a refusal of homogenizing integration.

For the modern state, time is conceived as linear, measurable, and subordinated to schedules, labor contracts, and administrative deadlines. For the Roma, by contrast, time is often rhythmic, attuned to natural cycles, family events, and cultural practices; it is not "managed" but experienced. This should not be perceived as a "deficit" of modernity but rather as an alternative ontology of temporality, one that enables Roma communities to operate within the contemporary world without fully adopting its normative premises.

Central to this autonomy is mobility, often misread by institutions as instability or antisocial behavior but in fact a cultural strategy of survival and adaptation, enabling Roma to evade control, escape persecution, and pursue precarious opportunities through kinship networks and ethical codes—a form of "tactical freedom" (de Certeau, 1984). Likewise, home is not defined by property or fixed structures but by community, kinship, ritual, and shared practices; even in ghettos or temporary shelters, Roma enact home through food, music, communication, and ritual, creating a cultural ontology of belonging beyond land deeds or legal registration.

Home in the presence of loved ones, the ritualized and communal space. It is a cultural ontology of belonging that does not require land deeds or address registration to be real. In this sense, even when Roma inhabit ghettos, temporary shelters, or semi-ruined buildings, they enact home through symbolic practices—assembling space, sharing food, making music, engaging in communication, performing rituals and so on.

Beyond commerce, the market for the Roma is a communal home-making space where exchange, presence, and ritual create a lived sense of belonging, reflecting their broader view of home as rooted in social bonds rather than physical structures. Home is above all a social and emotional configuration, with family and kinship at its core. Unlike the postmodern nuclear model, the Roma family is extended, hierarchical, and deeply embedded in the social fabric, providing emotional support while also serving as an institution of power, economy, justice, and education.

In this light, the "Straw Road" and the glass dome converge as metaphors for two worlds—the mobile and the institutional—that can coexist not through subordination but through harmonious attunement. This reveals the potential compatibility between cultures with differing temporalities, modalities, and spatial logics. The Roma trajectory should not be interpreted as a lack of history, but as a cultural critique of modernity, experienced through autonomous regimes of time, belonging, and cultural production.

Geography of Despair

The history of the ghetto does not begin with modern urbanization; it traces back to the sixteenth century with the emergence of Jewish quarters in Venice. These spaces were created not merely for isolation but as instruments of institutionalized control, subjugation, and surveillance (Sassen, 2014). This model has been repeatedly reproduced across different cultural and historical contexts—from Black neighborhoods in the United States after the era of slavery, to the French *banlieues*, and to informal Roma quarters in southeastern Europe (see Fig. 2 and Fig. 3 illustrating the scale of its impact).

In its contemporary form, the ghetto represents a secondary product of economic rationalization, real estate speculation, and the systematic neglect of the social dimension of space. As David Harvey (2012, p. 45) notes, "cities were built for profit, not for people." He further develops this argument by analyzing urbanization as a mechanism of capital accumulation through dispossession—a process in which certain groups are systematically deprived of access to resources, housing, and public services.

Investment flows are systematically directed toward privileged urban zones, while ghettos remain in a state of institutional neglect functioning as sources of cheap labor and politically passive electorates. This produces a dual urbanization:

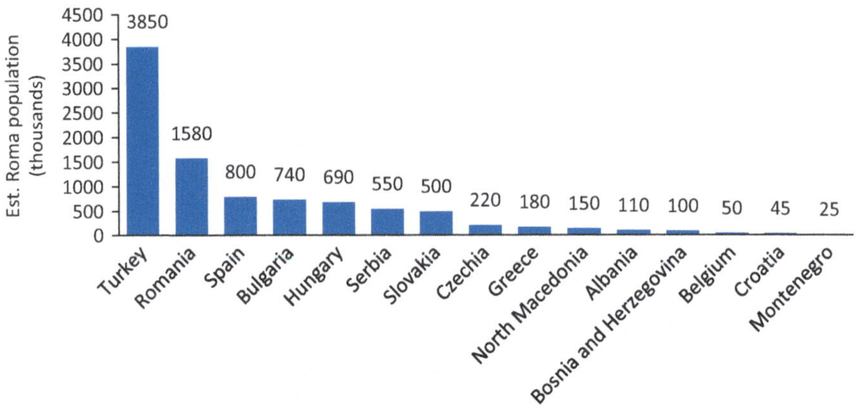

Fig. 2 Estimated Roma population (in thousands). [Values determined by percentage from the Council of Europe (2012) times the country population in 2023; these are not official statistics, but provisional estimates provided for analytical purposes]

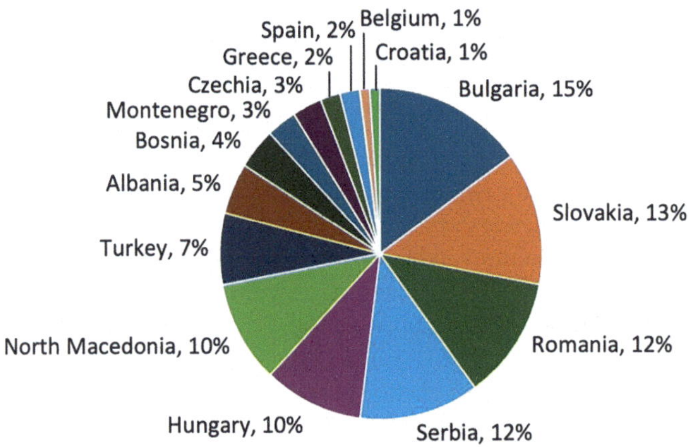

Fig. 3 Roma as a percentage of the total population in European countries. (Data from Strasbourg: Council of Europe, Roma and Travelers Division 2012. These figures should be treated as approximate, as censuses rarely provide reliable ethnic data. Among Roma populations, numbers are generally underestimated because many socially integrated Roma choose not to report their ethnic identity due to fear of stigmatization, exclusion, discrimination, or persecution. The expression of preferred, kaleidoscopic, or selective identities is an inherent aspect of Roma existence)

simultaneously coexisting yet radically unequal in terms of access, resources, and visibility—a "shining façade" and a "deep shadow" (Fig. 4).

Understanding the spatial formation of a Roma ghetto is essential to grasping both the constraints and strategies of everyday life. The patterns of density, segregation, and circulation shape social, economic, and cultural practices within the community. The Arman Mahala in Plovdiv formed over time, from 1952 to 2018, revealing the processes through which the ghetto's physical and social structure has developed.

Fig. 4 "The Shining Façade"—a residential building in the "White" part of Plovdiv, Bulgaria. In the parallel dimension—"*The Shadow*"—a residential apartment building in the largest Roma ghetto in Europe—"Stolipinovo," Plovdiv, Bulgaria. (Source: Krasimir Asenov)

The ghetto is not only a site of suffering but also of resistance. From the Watts riots in 1965 and Parisian suburb unrest in 2005 to contemporary anti-police movements in Latin America, it has served as a stage for collective uprising. For Fanon (2004, p. 23), revolt is existential, restoring dignity: "The oppressed discover themselves in the act of violence by which they break the chains." Resistance thus responds to systemic dehumanization while affirming human presence in an unrecognizing world.

The contemporary ghetto is shaped not only by racial or ethnic segregation but by capitalist-driven urbanization, where development serves market interests over social justice. Soja's (2010) concept of "spatial justice" highlights that when access to basic services depends on postal codes, cities fragment into "zones of life" and "zones of neglect," with the ghetto exemplifying the latter. It is both a product of neglect and an active social architecture where informal structures—gangs, religious groups, and parallel security—fill the state's absence. Under capitalism, not all bodies are "investable" (Mbembe, 2003, p. 25). Lefebvre's "right to the city" frames urban participation as fundamental, yet the ghetto denies this right, shaping space "to" its inhabitants rather than "by" them.

Brazilian *favelas*, Asian and Latin American *bidonvilles*, African settlements like Kibera, Momaka, and Mirwa, and Europe's largest Roma ghetto, Stolipinovo, are part of a global network of roughly 250,000 highly marginalized urban areas. On the fringes of hyper-urbanized cities such as Lagos, Kinshasa, Dar es Salaam, Delhi, and Karachi, these spaces face social conflict, infrastructure collapse, and public order breakdowns. Deprived of basic resources—water, electricity, healthcare, justice—life unfolds in a parallel reality of despair and improvised survival. The state is present mainly through repression, while paramilitary groups, drug cartels, and criminal networks fill the power vacuum. These ghettos are autonomous worlds, self-organizing under often brutal rules, reflecting the twenty-first-century urban archetype: cities increasingly resemble agglomerations of metal, insecurity, and alienation, producing dependency and self-aggression—the "syndrome of deprivation."

The parallel world of Roma communities in Europe exists not only as a cultural–historical phenomenon but also exhibits a distinct spatial logic—what critical geography terms the "geography of despair" (Wacquant, 2007). Roma embodiment and cultural memory are literally inscribed in marginalized zones—informal settlements, industrial ruins, and abandoned housing. These are not just symbols of poverty but "geosociocultural containers" of exclusion—spaces in which social death is lived and experienced (Bourdieu, 1999, p. 125).

The Syndrome of Deprivation: When the State Is Present Through Absence

The "syndrome of deprivation" is not a medical diagnosis or formal social science term, but a metaphor of deep theoretical and existential weight. It describes how entire communities are systematically stripped not only of resources but also of rights, identity, space, and a future. Produced by political decisions, urban practices, and long-term marginalization, it finds one of its clearest and most painful expressions in the everyday reality of Roma communities in Europe.

These settlements, understood both geographically and politically, operate as zones of exclusion where the state manifests less through support and more through repression—police raids, forced evictions, and administrative neglect.

This practice transforms the settlement into a modern camp in the sense of Giorgio Agamben who argues that "the camp is the nomos of modernity" (Agamben, 1998, p. 11), where such spaces are not exceptions but reflect the norm of the modern state. Here, Roma people become *homo sacer*—beings who can be killed but not sacrificed, as they possess neither political status nor legal protection (Agamben, 1998, p. 8), whose lives exist as "bare life," stripped of recognition, dignity, and political agency.

This dynamic aligns with Michel Foucault's (1976) concept of biopolitics, where power governs not by direct killing but by determining who may live or die through the systematic denial of resources like healthcare, education, and housing. In Roma settlements, this takes the form of structural neglect—absent sewage, water, electricity, schools, and medical services—reflecting political decisions that mark these communities as "unsuitable" for full social integration. The settlement thus becomes a biopolitical zone of abandonment where life is neither destroyed nor nurtured but left unsupported—a form of structural violence enacted through the absence of state care rather than overt repression.

Roma children, born into chronic poverty and denied education, are systematically deprived of a future, trapped in a cycle of marginalization. Among the most excluded and vulnerable in the EU and its partner countries, their condition stems not from individual or cultural shortcomings but from long-standing political practices of discrimination, spatial segregation, and institutional neglect.

According to European Commission (2021) data, over 50% of Roma children in Bulgaria, Romania, Serbia, and North Macedonia live in households experiencing severe material deprivation, compared to approximately 7% of the general population. Furthermore, only 15–25% of Roma children attend preschool institutions, in contrast with over 90% of non-marginalized peers (European Commission, 2020, EU Roma Survey).

This indicator measures the share of individuals avoiding state institutions due to fear of discrimination, rights violations, or inadequate treatment. Among Roma, the rate is 63%, reflecting institutional isolation—withdrawal not from lack of need but from perceiving institutions as threatening. This aligns with Bourdieu and Wacquant's (1999) concept of institutional violence, where legally sanctioned state practices—refusing to register children, police raids, forced evictions, medical neglect—exert control and produce trauma. Such experiences create collective warnings like "don't go to the police" or "don't trust the doctor," leading to self-isolation that restricts education, healthcare, social services, and legal protection—a cycle of social exclusion. By contrast, only 14% of the general population avoids institutions, reflecting ordinary skepticism rather than traumatic avoidance.

Together, these indicators show that the Roma in Europe face not just poverty or cultural lag, but persistent political and institutional exclusion maintained through:

- **Low trust** → resulting from lack of justice and political representation.
- **High sense of historical injustice** → stemming from unacknowledged trauma and systematic silencing.
- **Massive institutional avoidance** → driven by direct experiences of discrimination and violence.

These data highlight the urgent need for reparative policies—recognizing historical violations, reforming institutions, and restoring trust through meaningful participation. Such measures require addressing the long-standing symbolic and political invisibility of Roma populations, whose marginalization stems not from poverty but from deliberate strategies that erase them from the national narrative while casting them as a social "other" in political and media discourse.

Jean-Paul Sartre emphasizes that exclusion is not only material but existential, stripping freedom of choice and reducing individuals to survival-driven beings. As he notes, "You are what you do with what has been done to you" (Sartre, 2007, p. 54), underscoring that Roma are not inherently marginalized—centuries of slavery, discrimination, forced assimilation, and social displacement have rendered them so. In this context, Sartrean freedom becomes illusory, as choices are limited to survival rather than personal growth or social advancement.

Simone Weil calls this condition "extracted life"—the loss of identity and ability to act as a subject when severed from land, culture, and history (Weil, 2002, p. 89). Among Roma, it appears in the systematic destruction of language, traditions, and practices: Romani language is dismissed as a "dialect" or "jargon" and children are forced to speak only state languages, leading to cultural erasure. Clinically, this is "cultural bereavement" (Bhugra, 1998, p. 167)—a trauma marked by loss of identity, tongue, and social ties, producing depression, apathy, and aggression. Widespread in Roma communities, it remains largely unrecognized and untreated due to poor access to services and the denial of cultural trauma as a valid medical or social issue.

Yet the settlement is not only a site of suffering but also what Deleuze and Guattari (1987, p. 379) call a "smooth space," where power cannot fully dominate, allowing alternative life, resistance, and cultural creation. In Roma settlements, this space appears in strong community ties, mutual aid, improvised infrastructure, and resilient practices like music, dance, and rituals—active forms of resistance against erasure. As they note, "the ghetto can also be a machine for war—social, cultural, spiritual" (Deleuze & Guattari, 1987, p. 394).

Despite systemic exclusion, marginalized Roma communities sustain authentic subjectivity and alternative modes of being: Romani music, though often commercialized, remains a medium of self-definition and resistance, and the language endures in private and community spaces as a symbol of resilience. Yet this "micro-resistance" is limited by structural violence—forced evictions in Bulgaria, Romania, and Slovakia displace communities to peripheral areas lacking jobs, transport, and services, masking poverty rather than addressing it. European "integration" policies often demand assimilation, forcing Roma to renounce their culture, language, and identity, perpetuating marginalization and enacting a modern form of colonial violence that undermines Romani subjectivity.

Overcoming the syndrome of deprivation requires more than resource distribution; it demands recognizing Roma as full citizens by valuing their culture, language, and history while ensuring access to education, employment, political representation, and rootedness. As Simone Weil (2002, p. 89) notes, "to be torn from the world is a form of death"; restoring life thus means restoring ties to land, culture, and rights.

Roma settlements are not natural but products of political decisions, and their exclusion undermines democracy by denying full citizenship to part of the population. The syndrome of deprivation, therefore, signals not only a need for policy reform but also an ethical reevaluation of contemporary citizenship.

The Architecture of Exclusion: The Roma Mahala as a Spatial-Political Category

Urban space is never neutral; shaped by power, social conflict, and political decisions, architecture serves not only aesthetics or function but also control, segregation, and exclusion. Roma *mahalas* in Central and Eastern Europe are not mere outcomes of poverty or spontaneous urbanization but materially constituted spaces of systemic exclusion—products of institutional policies, neglect, and discrimination. As an "architecture of exclusion," these spaces do more than contain poverty: they reproduce the social and cultural subordination of Roma, regulating access, movement, visibility, and belonging. As Edward Soja notes, "the city is not only a place of life, but an instrument for isolating lives" (Soja, 2010, p. 89).

In Roma *mahalas*, absent streets, enclosing walls, broken bridges, and dense constructions create both physical and symbolic barriers, deliberately engineered to confine poverty to the periphery. The *mahala* thus operates as a political formation where state control is enforced through infrastructure, repression, police raids, evictions, and sanctions, functioning as "camps in peacetime" (Agamben, 1998, p. 9). Residents often lack documents, addresses, or state records, rendering them both invisible and constantly surveilled—a paradox of the architecture of exclusion. Labels like "chaotic" or "illegal" obscure the political reality resulting not from Roma "incompetence" but from deliberate state neglect.

This constitutes "deliberate inaction," an institutional strategy in which the state avoids investing in areas deemed "temporary," "illegal," or "unplanned," even after decades of existence. David Harvey (2008, p. 53) frames this as "spatial justice," where access to resources, services, and quality urban space is determined by class and ethnicity. In Roma *mahalas*, this injustice is extreme: residents occupy homes that can be demolished at any moment without legal recourse, creating constant insecurity—an architectural reflection of the legal vacuum in which they are abandoned. Agamben describes this as "encampment logic," a state of permanent temporariness lacking ownership, stability, or future prospects (Agamben, 1998, p. 11). This logic extends beyond refugee camps to Roma settlements, where even legally constructed homes are often demolished for "urban planning violations," despite official plans historically failing to recognize Roma as legitimate citizens.

The physical separation of Roma *mahalas* is enforced through walls, buffer zones, broken streets, poor transport, and impassable terrain. In Bulgaria, cities like Stara Zagora and Sofia use concrete walls under the guise of "order and security" to mark Roma as "other" and inscribe ethnic stigma. This isolation drives educational, economic, and political marginalization—children struggle to access schools, youth to find jobs, and adults to receive social assistance—naturalizing inequality and masking discrimination as cultural deficit.

The architecture of exclusion is an active system reinforcing marginalization through: (1) displacement to peripheral, often hazardous areas; (2) infrastructural neglect—absent water, sewage, electricity, and internet; (3) urban marginalization, with *mahalas* omitted from official plans and rendered "invisible"; and (4) architectural stigmatization, labeling layouts "chaotic" or "temporary" to justify eviction. Across Europe, this reflects Foucault's (1977, p. 202) "architecture of surveillance," with Roma settlements monitored via police patrols, video surveillance, and restrictive social services, limiting mobility, access to institutions, and maintaining constant threat of intervention.

The architecture of exclusion turns the *mahala* into a "black hole" of citizenship—where the state withholds services and rights yet remains predatory, demanding obedience and imposing penalties.

Roma exclusion extends beyond the spatial segregation of the *mahala* to the labor sphere, where economic participation is limited to informal survival strategies, with professions like street vending, waste collecting, and shoe-shining emblematic of stigmatized, precarious livelihoods (Fig. 5).

Survival occupations mark poverty and exclusion, creating a "parallel world" (Yiftachel, 2009) where Roma sustain their own organization, economy, and culture under threat of erasure. Yet resistance emerges even in oppressive spaces: through self-built homes, improvised plumbing, graffiti, music, and rituals—acts asserting presence, demanding recognition, and transforming survival into resistance.

Despite this resilience, the architecture of exclusion is legitimized through discourses of "order," "urbanism," and "integration." European programs often reduce forced evictions or relocation to homogeneous social housing, severing residents from networks and culture. Such interventions reshape rather than end exclusion, converting *mahalas* into socially engineered spaces while leaving Roma citizens' status unchanged. As Roy (2005, p. 147) notes, this denies them the "right to the city."

Fig. 5 Shoe shiners in parallel worlds: the last shoe shiner on the Balkans, a vanishing trade reflecting both marginalization and the dignity of labor in post-socialist society, and a shoe shiner in New York City, where the same humble profession persists in a global metropolis—evidencing both continuity and contrast in urban economies. (Source: Krasimir Asenov)

Overcoming exclusion requires dismantling the power relations that produce it: recognizing Roma as full participants in urban planning, granting land and property access, and enabling meaningful involvement in decisions about their space. Most urban projects are imposed, turning *mahalas* into sites of political erasure. The architecture of exclusion is thus a visible testament to moral failure, showing not all city residents enjoy a dignified existence.

Conclusion

The parallel world of the Roma shows that their existence, culture, and community cannot be fully understood through the lenses of state, law, or market alone. From the "Straw Road" to contemporary ghettos, Roma life reflects a complex interplay of autonomy, resilience, and cultural creativity. Their habitats—often framed as sites of exclusion—are also spaces of vibrant cultural production, where music, dance, color, and ritual convey both the pain of marginalization and the determination for symbolic survival. In these spaces, Roma do not simply navigate modernity; they inhabit and reshape it on their own terms, creating alternative forms of social organization, knowledge, and expression that challenge dominant notions of progress, rationality, and value.

This cultural autonomy exists alongside persistent material and political marginalization. Overcoming the architecture of exclusion requires dismantling the power relations that produce it, recognizing Roma as full participants in urban planning, granting land and property access, and empowering them to shape decisions about their own spaces. Only when treated as active subjects can architecture shift from exclusion to inclusion. Imposed interventions without participation turn the *mahala* into a site of political erasure, where life persists without recognition. At its core, the architecture of exclusion is a visible testament to moral failure, showing that not all urban residents enjoy the right to dignity.

Future research should examine survival mechanisms in contexts where other cultures would struggle, alongside the strong kinship bonds that shape the Roma cultural matrix. Central is their adaptability—not passive assimilation but the strategic use of dominant cultural environments as resources for self-strengthening. This transformation of the "foreign" into a tool of autonomy opens new fields of inquiry: from the ghetto as a creative habitat and economies of survival and mobility, to the anthropology of cultural transgression, where exclusion and self-construction blur to produce new forms of modernity.

Seeing the Roma as participants in a parallel world thus invites a rethinking of modernity. Ghettos are not just spaces of despair but laboratories of alternative social, cultural, and political life. Recognizing this reality highlights the ethical and practical need to engage difference, creativity, and autonomy in urban life. Through cultural agency and inventive adaptation, the Roma reveal possibilities for living otherwise, challenging us to reimagine cities, societies, and modernity in ways that honor human diversity, agency, and dignity.

References

Agamben, G. (1998). *Homo Sacer: Sovereign power and bare life* (D. Heller-Roazen, Trans.). Stanford University Press.

Agier, M. (2016). *Borderlands: Towards an anthropology of the condition of exile* (D. Fernbach, Trans.). Polity Press.

Assmann, J. (2011). *Cultural memory and early civilization: Writing, remembrance, and political imagination* (D. H. Wilson, Trans.). Cambridge University Press.

Assmann, J. (1995). Collective memory and cultural identity. *New German Critique, 65*, 125–133.

Bauman, Z. (2004). *Wasted lives: Modernity and its outcasts*. Polity Press.

Bhugra, D. (1998). Migration and mental health: An interface. *International Journal of Social Psychiatry, 44*(3), 167–173.

Bourdieu, P. (1999). *The weight of the world: Social suffering in contemporary society* (P. P. Ferguson et al., Trans.). Polity Press.

Bourdieu, P., & Wacquant, L. J. D. (1999). On the cunning of imperialist reason. *Theory, Culture & Society, 16*(1), 41–58.

Connerton, P. (1989). *How societies remember*. Cambridge University Press.

Council of Europe, Roma and Travellers Division. (2012). *Estimates on Roma population in European countries*. Council of Europe. http://rm.coe.int/CoERMPublicCommonSearchServices/DisplayDCTMContent?documentId=0900001680088ea9

de Certeau, M. (1984). *The practice of everyday life* (S. Rendall, Trans.). University of California Press.

Deleuze, G., & Guattari, F. (1986). *Nomadology: The war machine* (B. Massumi, Trans.). Semiotext(e). (Original work published 1983).

Deleuze, G., & Guattari, F. (1987). *A thousand plateaus: Capitalism and schizophrenia* (B. Massumi, Trans.). University of Minnesota Press.

European Commission. (2020). EU-MIDIS II—Survey results at a glance: The experience of Roma in 11 EU Member States.

European Commission. (2021). Council recommendation of 12 March 2021 on Roma equality, inclusion and participation (2021/C 93/01). *Official Journal of the European Union*. https://eur-lex.europa.eu/legal-content/EN/TXT/PDF/?uri=CELEX%3A32021H0319%2801%29

Fanon, F. (2004). *The wretched of the earth* (R. Philcox, Trans.). Grove Press. (Original work published 1961).

Fanon, F. (2008). *Black skin, white masks* (R. Philcox, Trans.). Grove Press. (Original work published 1952).

Foucault, M. (1976/2003). *Society must be defended: Lectures at the Collège de France, 1975–1976* (D. Macey, Trans.). Picador.

Foucault, M. (1977). *Discipline and punish: The birth of the prison* (A. Sheridan, Trans.). Pantheon Books.

FRA–European Union Agency for Fundamental Rights. (2022). *Roma in 10 European countries: Main results from the 2021 Fundamental Rights Survey*. Publications Office of the European Union.

Goody, J. (1987). *The interface between the written and the oral*. Cambridge University Press.

Graeber, D. (2011). *Debt: The first 5,000 years*. Melville House.

Harvey, D. (2008). The right to the city. *New Left Review, 53*, 23–40.

Harvey, D. (2012). *Rebel cities: From the right to the city to the urban revolution*. Verso.

Howes, D. (Ed.). (2003). *Sensual relations: Engaging the senses in culture and social theory*. University of Michigan Press.

Ilieva, N., Asenov, K., Kazakov, B., & Lubenov, T. (2020). Dynamics of the internal structure and habitation conditions of the ghettoized urban structure of Harman Mahala quarter—Plovdiv, Bulgaria (through combined application of remote sensing and field research methods). In: S. Nedkov et al. (Eds.), *Smart geography*. Key challenges in geography. Springer. https://doi.org/10.1007/978-3-030-28191-5_14

Ingold, T. (2011). *Being alive: Essays on movement, knowledge and description*. Routledge.
LaCapra, D. (2001). *Writing history, writing trauma*. Johns Hopkins University Press.
Lefebvre, H. (1991). *The production of space* (D. Nicholson-Smith, Trans.). Blackwell. (Original work published 1974).
Lucassen, L., Willems, W., & Cottaar, A. (1998). *Gypsies and other itinerant groups: A socio-historical approach*. Palgrave Macmillan.
Mbembe, A. (2003). Necropolitics. *Public Culture, 15*(1), 11–40.
Nancy, J.-L. (2000). *Being singular plural* (R. D. Richardson & A. E. Oustinoff, Trans.). Stanford University Press.
Ong, W. J. (1982). *Orality and literacy: The technologizing of the word*. Methuen.
Open Society Foundations. (2020). *Historical injustice and collective memory among Roma communities in central and Eastern Europe*. OSF.
Ricoeur, P. (2004). *Memory, history, forgetting* (K. Blamey & D. Pellauer, Trans.). University of Chicago Press.
Roy, A. (2005). Urban informality: Towards an epistemology of planning. *Journal of the American Planning Association, 71*(2), 147–158.
Sartre, J.-P. (2007). *Anti-semite and Jew* (G. J. Becker, Trans.). Schocken Books.
Sassen, S. (2014). *Expulsions: Brutality and complexity in the global economy*. Harvard University Press.
Scott, J. C. (2009). *The art of not being governed: An anarchist history of upland Southeast Asia*. Yale University Press.
Sigona, N., & Vermeersch, P. (2012). *Roma in Europe: A policy challenge*. COMPAS, University of Oxford.
Soja, E. W. (2010). *Seeking spatial justice*. University of Minnesota Press.
Van Baar, H. (2011). *The European Roma: Minority representation, memory and the limits of transnational governmentality* [Doctoral dissertation, Utrecht University].
Wacquant, L. (2007). Territorial stigmatization in the age of advanced marginality. *Thesis Eleven, 91*(1). https://doi.org/10.1177/0725513607082003
Warren, M. E. (2006). *Democracy and trust*. Cambridge University Press.
Weil, S. (2002). *The need for roots: Prelude to a declaration of duties towards mankind* (A. Wills, Trans.). Routledge.
Williams, R. (1977). *Marxism and literature*. Oxford University Press.
Yiftachel, O. (2009). Critical theory and 'gray space': Mobilization of the colonized. *City, 13*(2–3), 240–256.

Kazakhs and Russians in Kazakhstan: Divided/United We Stand…Still

Nygmet Ibadildin

Introduction

Kazakhstan is a country shaped by multiple layers of history—from ancient times to the present day: Scythian, Turkic, Genghis Khan, Kazakh khanates, Russian imperial, Soviet, and independent periods. Its current ethnic diversity, often framed as an achievement of peaceful coexistence, is also the result of forced migrations, deportations, colonization, political exile and engineering. Among the many communities that make up Kazakhstan, the two most prominent—Kazakhs and Russians—have long lived side by side, sharing the same cities and institutions, but often experiencing the country through very different lenses.

This research examines how Kazakhs and Russians continue to live in what can be described as parallel universes—a metaphor that reflects differences not only in language, religion, or culture, but also in historical memory, mindsets, media consumption, and even geopolitical orientation. These differences are not always visible on the surface. Daily life in Kazakhstan offers many examples of integration, cooperation, and shared space. But underneath, there are also tensions, silences, and competing narratives that shape how people see the country—and each other.

First, we look at the history and background of how two worlds emerged and co-exist. Next, we consider power—how it was distributed historically in the Soviet Union, how it has shifted after independence, and how this continues to produce both privilege and resentment. In more subtle ways, these divides are also visible in jokes, songs, memes, and online commentary, which express both frustration and belonging.

N. Ibadildin (✉)
Department of Political Science and International Relations, College of Social Sciences, KIMEP University, Almaty, Kazakhstan
e-mail: nygmet@kimep.kz

S. D. Brunn (ed.), *Transdisciplinary Research Across Knowledge Worlds*,
https://doi.org/10.1007/978-3-032-19491-6_6

The question is whether these parallel universes will continue to grow further apart, or whether shared challenges will bring people together to build a common and safe environment of trust, joint habitation, and cooperation. This paper does not claim to offer final answers. It is an attempt to map the space in between—to capture how history, identity, and politics create different realities within the same borders. Kazakhstan is not unique in this position, as many post-colonial and post-imperial states deal with similar questions. But in our case, the dividing lines are close, sometimes invisible, and still very much alive.

History and Memory: Origins, Legacies, and the Contested Past

Kazakh states or khanates—while nomadic—effectively controlled territory, codified social relations and property, and had power institutions and international recognition as seen in diplomatic correspondence with neighboring states. Russian and Western researchers used to label these states as tribal unions, tribes, or nomads implying a certain inferiority. But these states were effective enough to control their territory. However, they missed the Industrial Revolution and eventually became objects of colonial expansion by a more powerful state—Tsarist Russia.

Russian expansion was not successful until the mid-nineteenth century as Russians struggled to overcome the tyranny of space and faced logistical challenges in controlling the vast steppes. Kazakhs resisted effectively, using the advantages of their enormous territories. Eventually, Russian authorities constructed a line of fortresses where troops could access water and rest, thereby facilitating further expansion. By the end of the nineteenth century, they had taken control of Kazakh lands and declared them the property of the Russian Tsar. Kazakh khanates were abolished by Russian decrees. Kazakhs were expelled from their lands and then given to Cossacks—the vanguard of Russian imperialism—and to Russian peasants who settled in Kazakhstan under colonial protection. The first Russians who came to Kazakhstan were those who took advantage of the military conquest.

Land rivalry led to numerous rebellions and wars. Some Kazakhs chose to support Russian authorities believing it was necessary to adopt advanced technologies and knowledge from them. The 1868 *Steppe Statute* formalized the theft of land; Kazakh pastures were given to Cossacks and Russian settlers while local rebellions were crushed (Martin, 2001). The 1916 uprising demonstrated both Kazakh resistance and the brutality of reprisals—approximately number of Kazakhs and Kyrgyz fell from 937,000 to 670,000 from 1916 to 1917 (Pianciola, 2019a, p. 169). As Pianciolla (2019a) continues "Many fled to China, but the extermination targeting Kyrgyz and Kazakh men, women and children claimed a number of victims that most probably reached the tens of thousands". In interview an editor of the volume *The Central Asian revolt of 1916. A Collapsing Empire in the Age of War and Revolution* Alexander Morrison (2021) mentions the figure of 150,000 human loss for Zhetysu region where Kazakhs and Kyrgyz were main participants of the rebellion. Marcko Buttino gives estimates from 270,000 to 150,000 human loss among Kazakhs and Kyrgyz in Zhetysu region (Buttino, 2003, p. 80). We have to understand that the rebellion was in many places in Central Asia including Northern and Central Kazakhstan, which is not Zhetysu and Turkmenistan and Uzbekistan, so the losses can be much higher.

The 1916 uprising in Central Asia led to massacres on both sides—Kazakh and Russian—and was never fully suppressed as the February Revolution and October coup took place in Russia in 1917.

Kazakh intellectuals attempted to declare autonomy and establish an independent Kazakh Alash Orda Republic, but the Russian Civil War was unfolding actively in Kazakhstan where Russians were numerous. Kazakhs were divided during the civil war. Some supported the Whites, others the Reds, and many tried to stay out of the conflict altogether. When Alash Orda was planning nation-building and creating key institutions, it intended that Russians would receive proportional representation.

The Alash government's 1919 constitution proposed a modern Kazakh state with land reform and women's rights (Kendirbai, 2002, p. 14). However, their convocation lasted only a few months as the government was caught between White generals who rejected autonomy and Bolsheviks who falsely promised it (Payne, 2011). When Alash leaders negotiated autonomy (1917–1920), they faced impossible choices. As Kendirbai (2002) acknowledges, their temporary alliance with the Bolsheviks was not ideological but pragmatic since the Whites offered only continued colonial subjugation. The current boundaries largely reflect that historical divide which was shaped through negotiations between local Central Asian elites and grassroots actors including Kazakhs, Uzbeks, Tajiks, Kyrgyz, and Turkmen as well as Moscow authorities.

The first decrees in Soviet Kazakhstan focused on regulating marriage customs: arranged marriages, polygamy, and bride payments were prohibited. Later, the historic self-designation of "Kazakh" was officially restored as the Russian colonial administration had previously labeled them "Kyrgyz." Land was returned to Kazakhs, and the migration of Russian settlers was partially restricted. This period was relatively favorable for Kazakhs. As Pianciola (2019b) observed:

> …Russian and Kazakh Bolsheviks implemented policies of hard decolonization (1921–1922): tens of thousands of Slavic settlers were expropriated, and land was redistributed to nomads. During the period 1923–1927, soft decolonization prevailed: Kazakhstan was established as an ethnonational administrative region, and agricultural immigration was prohibited. Kazakhs were prioritized in access to land and water and were included in state and party administrations.

However, Soviet liberalization and decolonization ended abruptly in 1928. Kazakh households were labeled as *bai* (wealthy) and the forced expropriation of livestock began. This policy was followed by the collectivization campaign (*Kollektivizatsia*) during which the majority of households lost their livestock. It led to an unprecedented famine and massive human loss. This period is remembered as *Asharshylyq* in Kazakh collective memory. Sarah Cameron (2018) argues that the 1930s were not merely about collectivization, but rather constituted "demographic warfare." Her research confirms that Kazakhs lost 38% of their population; 1.3 million people starved as livestock were seized. Simultaneously, specialists who conducted the 1937 census were purged and the actual number of victims remains contested.

It appears that nomadic Kazakhs were inherently difficult for the Soviet totalitarian state to control as they possessed their own food systems, shelters, transportation, social structures, culture, and language. The Soviet sedentarization campaign (1928–1933) systematically dismantled Kazakh nomadic pastoralism through livestock confiscation and enforced famine. Cameron (2018, p. 5) calls this "demographic warfare," which not only claimed lives but also destroyed the material basis of Kazakh cultural reproduction.

Stalin later repopulated the steppes with deported minorities (Polian, 2004), creating artificial diversity. These included Russian and Ukrainian *kulaks*, and later entire ethnic groups such as Koreans, Germans, Poles, Chechens, Balkars, Crimean Tatars, and many others. All were relocated to Kazakhstan as punishment for perceived disloyalty. Kazakhstan thus became a multiethnic republic with Kazakhs themselves eventually becoming a minority in their own homeland.

Industrial development in Kazakhstan accelerated partially during WWII as factories were evacuated from western Soviet territories, and again during the 1960s and 1970s. While some scholars (e.g., Khalid, 2021) emphasize the benefits of Soviet modernization, the demographic and cultural costs for Kazakh autonomy were catastrophic. Russification intensified in the late 1950s with the Virgin Lands Campaign (*Tselina*), which prioritized grain production in rural areas at the expense of traditional Kazakh pastoralism.

Independence and Consequences for Demography, Society, and Economy

This extensive historical background is necessary to understand the parallel and often unequal coexistence between Russians and Kazakhs in Kazakhstan today. It was not a typical dynamic of colonizers versus colonized in a classical empire as both groups experienced repression under the Tsarist and later Soviet regimes. Nevertheless, systemic privileges were clearly skewed in favor of Russians.

At the time of *perestroika*, Kazakhstan declared independence in 1991 as a country without a clear ethnic majority. Martha Olcott (1993) referred to it as a "republic of minorities." The primary challenge was to build a civic nation without ethnic dominance or resentment rooted in colonial trauma. As Dave (2007) documents, the legacy of Soviet Russification, particularly through campaigns like Virgin Lands, left Russian-speaking urban elites in dominant positions while rural Kazakhs remained marginalized.

What Kazakhstan inherited was a deeply fragmented society shaped by the Soviet experiment in forced unity, mass resettlements, and decades of Russification.

While Soviet modernization created an industrial and urban elite, it also disrupted traditional Kazakh life, marginalized rural communities, and left behind a complex linguistic and ethnic landscape. The result was a country where Russian-speaking urban Kazakhs and ethnic Russians often held cultural and social advantages while rural Kazakhs and deported minorities were relegated to the periphery.

We next discuss how these historically, socially, and economically divided worlds evolved and how they now interact—not to build walls, but to construct bridges.

Demography

Russians began actively leaving Kazakhstan in the early 1990s, a trend that continues to the present. At the same time, ethnic Kazakhs who had fled during the 1930s to China, Uzbekistan, Afghanistan, and other neighboring regions began returning to Kazakhstan. This return migration was supported as a state policy to address historical injustice. Another significant wave of Russian migration occurred in 2022 when Russia declared mobilization for the war in Ukraine. Many Russians fled to other countries, and Kazakhstan became an attractive destination due to its visa-free entry and widespread use of the Russian language.

The reasons behind the emigration of Russians from Kazakhstan since 1991 are multifaceted and remain open to debate. Various push and pull factors—economic, political, and psychological—can help explain this phenomenon (Auyezbek, 2023; Chulanova, 2021):

- Fear of ethnic pressures from Kazakhs, such as the requirement to learn the Kazakh language and a perceived loss of the privileges Russians enjoyed during the Soviet era;
- Perception of declining educational standards;
- Limited opportunities for themselves and their children;
- Better social and employment conditions in Russia;
- Emotional attachment to Russia as a homeland;
- Incentives offered by Russia for repatriation, including material support; and
- A combination of all the above.

Between 1989 and 1999, Kazakhstan's population declined by 9.1%, or more than 1.5 million people (Chulanova, 2021), primarily due to the emigration of ethnic Russians (Fig. 1).

Meanwhile, Russia itself has been facing a severe demographic crisis, often referred to as the "Russian cross," characterized by low birth rates, high death rates, and low life expectancy (Vishnevsky & Shkolnikov, 1997). At the same time, Russia has been interested in attracting migrants who speak Russian and share basic cultural values. Demographic patterns between Russians and Kazakhs in Kazakhstan differ significantly in terms of median life expectancy, death rates, fertility rates, and age composition. Russians tend to be older, with higher death rates, lower

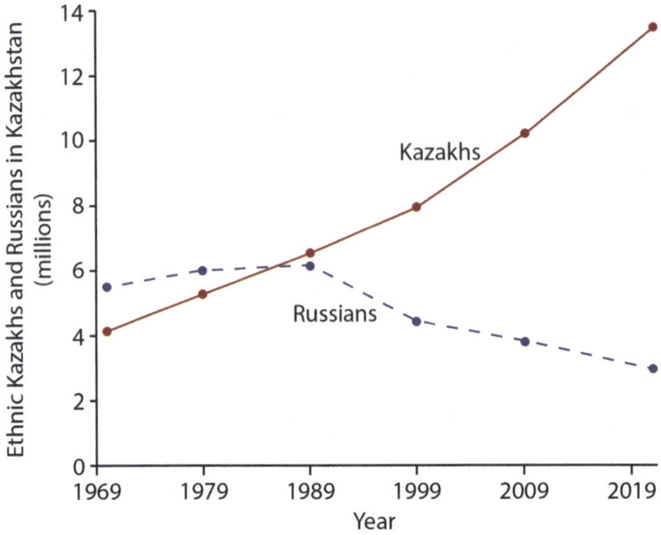

Fig. 1 Number of Kazakhs and Russians in millions from 1970 to 2021. (Census, 1974; Agency of the Republic of Kazakhstan for Strategic Planning and Reforms, 2023)

fertility rates, and a larger proportion of elderly individuals. Young Russians often leave Kazakhstan for education in Russia, which frequently leads to "delayed emigration" where parents eventually follow their children. However, Russian emigration has gradually declined since those willing to adapt to new realities chose to stay and tie their futures to Kazakhstan. The situation changed again in 2022 when Russia's mobilization announcement in response to heavy losses in the war against Ukraine prompted another wave of Russian migration to Kazakhstan as one of the most accessible destinations.

Geography and Memory Politics

Kazakhstan's physical geography consists of steppes, deserts, and mountains. Russian colonial expansion began in the north with the construction of fortresses designed to control territory and disrupt Kazakh nomadic routes (Ibadildin, 2024). These fortresses were manned by Russian Cossacks, a privileged military estate in the Russian Empire. Later, large waves of Russian and Ukrainian peasant settlers arrived who were protected by colonial authorities.

As a result of these historical developments, Russians are now concentrated in cities throughout Kazakhstan and especially in the northern and eastern regions. These areas suffered significantly during the man-made famine of the 1930s and were the primary targets of the Virgin Lands campaign which aimed to transform the steppe into a grain-producing region. Today, the northern and eastern regions of

Kazakhstan have Russian populations of up to 30%, and in some localities, even 50%. Kazakhs dominate demographically the southern and western parts of the country.

This geographic and historical settlement pattern has fueled irredentist rhetoric in Russia. Prominent figures like Aleksandr Dugin have referred to Northern Kazakhstan as "Southern Siberia" or "Internal East" (Dugin, 1997, p. 54). Since the 1990s, some Russian analysts have warned that a Russian "withdrawal from Central Asia" might lead to instability or violence. Scholars such as Zatulin and Migranyan (1997) have advocated for Russian intervention if anti-Russian sentiments rise in the region. While such views may not represent official policy, the full-scale invasion of Ukraine has made such scenarios seem more plausible.

Several Russian State Duma members—including Nikonov, Gurulyov, Fedorov, and Zatulin—have made inflammatory remarks about Kazakhstan's nation-building efforts and territorial integrity. Fears of irredentism in Northern Kazakhstan are real. German media published archived material in which a German diplomat quotes Putin in 1994, when he was an official in Saint Petersburg, claiming Crimea, Eastern Ukraine, and Northern Kazakhstan (Oshuev, 2025). At the time, he was an ordinary mid-level official, but even then he may have expressed the sentiments shared by many Russian officials.

The relocation of Kazakhstan's capital from Almaty to Astana (now again Astana) in 1997 may have been partially motivated by concerns over future instability in the north. One early irredentist episode occurred in 1991 even before the formal dissolution of the USSR. In Western Kazakhstan, historically a Kazakh-majority region but also a site of early Cossack fortifications, a group of Cossacks attempted to organize a convocation on September 15, 1991, potentially to call for annexation by Russia. The convocation, tied to the 400th anniversary of Cossack service to the Russian Tsar, raised fears of secession. It was peacefully disrupted by a massive Kazakh protest (Karabek, 2011). Russian media have since expressed regret over the failure of this effort.

In terms of memory politics and spatial representation, the Kazakh government initiated the renaming of cities, villages, and streets through popular vote, replacing colonial and Soviet names with Kazakh ones. This has often provoked resistance from local Russians. Many cities still retain colonial-era names, but renaming efforts are most contentious in large cities such as Petropavlovsk and Pavlodar named after Russian Tsars. Kazakh activists call for the restoration of their historical or indigenous names. Some compromise has been found in cities now recognized by both names or whose pronunciation varies between Russian and Kazakh—for example, Oral/Uralsk, Semey/Semipalatinsk, and Oskemen/Ust-Kamenogorsk. The city of Guryev was renamed Atyrau in 1992 with little public opposition.

Kazakh authorities began removing Soviet- and Russian-era monuments as early as the 1990s (Fig. 2). In fact, the very first monument taken down was that of Yermak, the famous Russian Cossack leader and brutal Siberian colonizer. The monument had been erected in the city of Yermak. When the city was renamed Aksu, the monument was removed. Interestingly, it was originally erected in Soviet

Fig. 2 Soviet monuments collected in a Kazakh amusement park. (Photo by the author, 2025)

Kazakhstan in 1965, then removed in independent Kazakhstan in 1992 and after its removal ethnic Russians transported the remnants to Russia and re-erected it in the Siberian city of Zmeinogorsk in 2006 (Romanov & Shchetinin, 2008).

The most widespread monument in the USSR was of Lenin. He stood in every city and probably in every village, usually in front of the main square or an official building. Almost all Lenin and other communist monuments in Kazakhstan were removed and stored as historical evidence in remote parks and alleys during the early 1990s. A rather unusual "removal" situation occurred in the city of Ridder, which was called Leninogorsk until 2002. Lenin's monument was relocated to a remote alley and the central location previously occupied by that monument was designated for a new monument to Philipp Ridder, a Russian general and geologist from the eighteenth century who likely founded the city and discovered its metallic ores. However, people began protesting due to the colonial connotations of such a monument, and authorities decided not to erect it even though it had already been cast. As of 2024, the space remains vacant (Bolatuly, 2024).

In sum, Russia's colonial expansion into Northern Kazakhstan in the nineteenth and early-twentieth centuries was later reinforced by the demographic catastrophe of the Soviet era, particularly the famine among Kazakhs. Today, this expansion has resulted in a complex and sensitive border region within contemporary Kazakhstan.

The country now faces the challenge of managing this legacy, seeking stability not through confrontation, but through careful administrative control and symbolic nation-building, that is, focusing on education, naming, language, and the encouragement of internal migration while at the same time avoiding the pitfalls of ethnonationalism. Yet, this quiet effort to assert sovereignty continues under the shadow of growing irredentist and revanchist rhetoric from Russia, which still frames parts of Kazakhstan as historically Russian territory.

Society and Politics

If we examine political and social protests, Kazakhstan is an authoritarian state where the consolidation of authoritarianism occurred gradually. In fact, most protests and social unrest in Kazakhstan are class-based, targeting inequality, the authoritarian nature of the state, and poverty. Over the past 30 years, a few ethnic clashes have occurred involving Kurds, Dungans, Armenians, and Chechens, usually stemming from criminal incidents and violence (Goble, 2021).

We can cautiously acknowledge that the Russian community has generally supported the Kazakh authorities, especially during the time of Nazarbayev, the republic's first president, and also the status quo (Lillis, 2014). One of Kazakhstan's most well-known journalists, Sergey Duvanov (2023), described the situation as follows:

> Other ethnic groups prefer not to get involved in politics or criticize the authorities. And this is the first indicator that Kazakhstan is a Kazakh state. I would describe the reserved attitude of most citizens from other ethnic groups toward the country's political life as a 'tenant syndrome.'

In a relatively soft authoritarian regime with open borders, those who are dissatisfied have the option to protest by emigrating elsewhere—and for many, Russia was the first choice. It could be said that most Russians who wanted to leave Kazakhstan have already done so. The lowest level of emigration from Kazakhstan was recorded in 2024 (Ranking.kz, 2025). A deep divide emerged in Kazakhstan after Russia invaded Ukraine in 2014 and even more so during the full-scale invasion in 2022. The events of 2022 were far more divisive due to the brutality of the war, including the bombing of civilians and the torture of prisoners and non-combatants by Russian forces (Gall, 2022). The Russian opposition media outlet *Meduza.io* published an article in 2014 about Russians living in Eastern Kazakhstan and their views on their position in the republic. Interviewees expressed frustration about renaming, the Kazakh language, the Kazakh population, discrimination, and drew comparisons between their situation and Crimea (Azar, 2014). The article touched on sensitive issues, and in the end, respondents who avoided discussing separatism still suggested that there were "enough people for the occasion."

> Despairing of finding even a single practicing separatist in Ust-Kamenogorsk, I ask Ataman Chmutov, 'Are there even many Cossacks around here?'
>
> 'Enough for the occasion,' he replies evasively. (Azar, 2014)

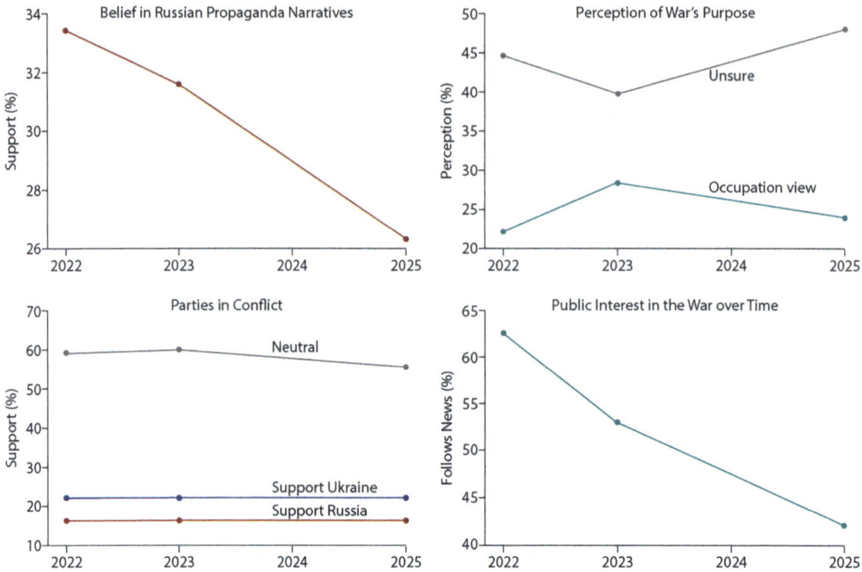

Fig. 3 Demoscope (2025) survey results on Kazakhstani opinions about Russia's invasion of Ukraine

After this article was published, *Meduza.io* was blocked in Kazakhstan and is now accessible only via VPN (virtual private network), which is commonly used in authoritarian countries to bypass government blocks of certain websites. Use of VPNs is currently prohibited in Russia.

Regarding the Russian invasion of Ukraine, public opinion in Kazakhstan remains divided: some support Russia, others support Ukraine, and many say, "this is not our war" (Fig. 3). As of 2025, around 52% of people in Kazakhstan are neutral, 15% support Russia, and 20% support Ukraine (Demoscope, 2025). The general trend is one of war fatigue: fewer donations are being sent to Ukraine now compared to 2022 when humanitarian aid from Kazakhstan—measured in tons—was flown in via airplane. Ukrainian journalists reported via a Telegram channel that 661 Kazakh citizens have fought as mercenaries for Russia, with 78 of them confirmed killed (Hochu_zhyt, 2024).

Economy

The Soviet legacy is a contradictory one. On the one hand, it proclaimed the right to self-determination and the emancipation of oppressed peoples. On the other hand, it imposed a far more severe system of oppression, particularly through the dominance of the Russian language and culture (Hirsch, 2005). In Kazakhstan, Russian populations predominantly live in urban centers while Kazakhs were concentrated in rural areas. Many historians now claim that Kazakhs were unofficially prohibited from living in cities as part of discriminatory policies (Bissenova, 2019). Today, poverty in Kazakhstan cannot be strictly categorized along ethnic or racial lines. However, some Kazakh journalists argue that Kazakhs remain disproportionately impoverished because they continue to live in rural areas and often lack inherited property or assets comparable to Russian families who typically own real estate in urban areas. Furthermore, the continued dominance of the Russian language in business creates barriers for Kazakhs seeking better-paid jobs (Darimbet, 2017).

One recurring economic and historical argument between pro-Russian and pro-Kazakh activists is the notion of "we built it for you" (Äliakbar, 2019, Nurmukhanbetov, 2022). This argument suggests that much of Kazakhstan's contemporary infrastructure—roads, schools, hospitals—was built during the Soviet era by Russians, implying that Kazakhs are now merely benefiting from these "gifts" (Exclusive.kz, 2025). Historically, however, Russian fortresses were constructed to control nomadic routes and water sources, thereby interrupting Kazakh seasonal movements and enabling colonization. Cities functioned as instruments of subjugation. Since most Kazakhs lived in rural areas, moving to a city was seen as a rare opportunity. Soviet policies until the early 1970s generally restricted internal migration by denying residence documents (*propiska*) to rural dwellers. This argument is not limited to social media or internet forums; it is often repeated in Russian public discourse as well. Some Russian emigrants have claimed that Kazakhstan's economy would collapse without Russians, portraying themselves as the engineers, managers, and educators who brought "civilization" to the so-called "barbaric" Kazakhs. Such narratives—about the loss of Soviet-era qualifications, historical memory, place names, and territory—often invoke the colonial trope: *we built it for you, we civilized you.*

Language Battles and Laws

The Kazakh and Russian languages have become highly politicized and even weaponized in Kazakhstan. In the 1990s, Kazakh was declared the state language, while Russian was designated a language of official use. This issue was so important that it was included in the Constitution. Article 7 of the Constitution of the Republic of Kazakhstan (enacted August 30, 1995) specifically states (see Akorda.kz, 1995):

1. The state language of the Republic of Kazakhstan shall be the Kazakh language.
2. In state organizations and local self-government bodies, the Russian language shall be officially used on an equal footing with the Kazakh language.
3. The state shall create conditions for the study and development of the languages of the people of Kazakhstan.

During the Soviet period, the Russian language held a privileged position while Kazakh was poorly taught in schools. Russian was the language of power, business, and education. Kazakh was mostly spoken in rural areas and associated with lower socio-economic status. As Kazakhs were not even the majority of the population at the time, this dominance of Russian continued largely without objection. The situation began to shift after Russia's invasion of Ukraine, which coincided with Kazakhs becoming the majority in Kazakhstan. More Kazakhs started demanding the use of the Kazakh language in public spaces and businesses such as cafés and restaurants. This led to immediate conflicts and mutual accusations. Kazakhs were frustrated that they could not receive services in their own language in their own country. Some began boycotting cafés that offered menus and service only in Russian.

Like many authoritarian regimes, Kazakhstan severely restricts freedom of speech. A special article was added to the Criminal Code to punish hate speech, but it has also been used against individuals who made critical statements about both Russian and Kazakh languages and ethnic groups. Here is the language in the article from the *Criminal Code of the Republic of Kazakhstan (2014)*:

> Deliberate actions aimed at inciting social, national, tribal, racial, class, or religious hatred; insulting national honor and dignity or the religious feelings of citizens; as well as propaganda of the exclusivity, superiority, or inferiority of citizens based on their attitude toward religion, class, national, tribal, or racial affiliation—if such acts are committed publicly, via mass media, telecommunications networks, online platforms, or through the production or dissemination of literature or other materials promoting social, national, tribal, racial, class, or religious hatred…

This article has been heavily criticized for limiting freedom of speech. Some activists and legal experts argue it should be repealed or amended to carry only administrative or civil penalties rather than criminal ones, as it suppresses open critique and dissent in society. The public discussion centers on the need to differentiate between hate speech, incitement to violence, or calls for foreign invasion, legitimate critique or disagreement. Bilingualism remains common in large cities, and non-Russians are usually fluent in both Kazakh and Russian. Officially, both languages are used in public and government communications (Fig. 4).

The celebration of holidays remains divisive not only for religious observances but also state commemorations. Russia has developed an almost religious cult around the "Great Patriotic War" (part of WWII). Kazakhstan's authorities and public are still grappling with how to frame and celebrate it. The President of Kazakhstan regularly attends the Russian military parade in Moscow organized by Putin. However, he does not wear the Russian military victory symbol—*the St. George's Ribbon (Georgievskaya lenta)*—which in Kazakhstan is seen as a colonial symbol dating back to the nineteenth century. This ribbon is now worn by pro-Russian military groups and the Russian army in its war with Ukraine. Notably, Russian soldiers and officers who participated in WWII also appeared in the 2025 parade. Over the past 15 years, Russia has introduced a civilian commemorative march known as *Bessmertny Polk* ("Immortal Regiment") where people carry portraits of relatives who fought in WWII alongside red flags and other Soviet symbols. These events have been held in Russian cities and were repeated several times in Kazakhstan. The author attended one for research purposes before 2022. It resembled a carnival, with people wearing *Star Wars* costumes and ironic slogans on their T-shirts.

Fig. 4 Public announcement in the park about dogs and grass in two languages. (Photo by the author, 2025)

After 2022 these actions were widely criticized on Kazakh social media as resembling support for Russia's war in Ukraine. While authorities banned such marches, some still took place. In response, Kazakhs participated by carrying national flags and singing the national anthem, a sharp contrast to the Russian style of commemoration. In 2025, one such event took place in Almaty, Kazakhstan's largest city, but its legal status was uncertain until the very day it took place (Fig. 5). There is a high probability that participants were paid by Russian sources (Exclusive.kz, 2025).

The controversy surrounding holidays and commemorations also reflects the deep influence of Russian media, TV, internet, and publishing on Kazakhstan. The country does not block the broadcasting of Russian propaganda, which has become a hot topic of debate. Fiction and non-fiction books by Russian authors promoting their narrative of victory in Ukraine are freely available in Kazakhstan. To be fair, however, books banned in Russia—such as biographies of Navalny or works by Akunin and Khodorkovsky—are also sold openly in Kazakhstan.

Language-related controversies and the "canceling" of public figures for disrespectful comments about the Kazakh language or people, sometimes with criminal consequences, are widely discussed on social networks and periodically addressed

Fig. 5 Public banner for commemoration of WWII victory in two languages. (Photo by the author, 2025)

by the authorities, including the President and with strong negativity from the Russian commentators in Russia.

Mixes in Flux: Visible and Invisible Streets, Food, and Schools

The Kazakh Constitution of 1995 declared equality for everyone and referred to a "peaceful civil society" with no reference to the ethnicity of Kazakh people. It is an obvious difference from the Constitution of 1993 and the declarations of state sovereignty of 1990 and state independence of 1991 where Kazakh people were referred to as nation-building ones.

The Declaration of State Sovereignty on October 25, 1990 was the earliest step toward independence, still in USSR times. It declared (Verkhovny Sovet of the Kazakh SSR, 1990):

> The revival and development of the unique culture, traditions, and language, as well as the strengthening of the national dignity of the Kazakh nation and other nationalities living in Kazakhstan, are among the most important tasks of the statehood of the Kazakh SSR.

The Declaration of Independence of December 16, 1991 (Verkhovny Sovet of the Republic of Kazakhstan, 1991) was officially adopted after the unsuccessful August coup in Moscow and the Almaty declaration in December 1991 when the USSR officially ceased to exist declared independence with a similar reference:

> The citizens of the Republic of all nationalities, united by a common historical destiny with the Kazakh nation, together constitute the single people of Kazakhstan, who are the sole bearer of sovereignty and the source of state power in the Republic of Kazakhstan, and who exercise state power both directly and through the state bodies elected by them on the basis of the Constitution and laws of the Republic of Kazakhstan.

The first Constitution of independent Kazakhstan, adopted in 1993, declared (Verkhovnyy Sovet Respubliki Kazakhstan, 1993):

> The Republic of Kazakhstan, as a form of statehood of the self-determined Kazakh nation, guarantees equal rights to all its citizens.

As we see, the current Constitution of 1995 avoids referring only to the Kazakh nation and ethnicity and makes an effort toward civil identity in nation-building. Multiethnicity is a common ground for society. Despite all post-colonial pains and resentments, everyday life does not have too many conflicts. Social networks are much more aggressive in this respect. Every-day coexistence between Kazakhs and Russians is peaceful and tolerant enough for Russians to stay, work, and raise children. Some regions have almost no Russians, and some have gained Kazakh people in the last 20 years. Looking at streets, cafés, bazaars, malls, schools, concerts, sports, and other public spaces—everything is ethnically mixed. Public transportation and taxis are mixed; drivers and passengers are united in scolding the government for inflation and prices. No one cares much who sits where or who drives whom. People live in the same apartment blocks, walk the same streets, and study in the same schools. Ethnic neighborhoods or villages of Kazakhs, Turks, Kurds, Uighurs, Russians, Germans, or Uzbeks historically exist but without clear boundaries.

The number of Russian schools is decreasing while Kazakh schools are growing in numbers, but Kazakh children are also more numerous now. In Soviet times, the number of Kazakh schools was small—Almaty, the largest city with 1 million people at that time—had only one Kazakh-language school. Intermarriages are also quite common, especially among secular or less conservative people. There is no taboo around it socially though families may sometimes prefer spouses from the same ethnic or religious background. Statistically, 18% of marriages in Kazakhstan in 2024 were inter-ethnic (Tarenchikova, 2025). The place of marriage is usually neutral—not mosque or church—but civil registry offices. Some Russian Muslims exist, just as some Kazakhs are Orthodox or Protestant Christians. Interestingly, the Russian Orthodox Church now even holds sermons in the Kazakh language, and the Bible has long been translated into Kazakh. One Russian Orthodox priest, Iakov Vorontsov, who was later fired from the church, called for establishing a separate Kazakh Orthodox Church without subordination to the Moscow Patriarchate and started collecting electronic signatures via the government petition portal (Almas, 2025).

Museums are a bit of a stub, as the Soviet period is not well represented with all its contradictions—Great Terror on one side and industrial development on the other. They mainly focus on pre-Russian nomadic history and post-independence periods and try to highlight the contributions of all ethnic groups to Kazakhstan's well-being. But they avoid articulating colonial violence or Russian atrocities. The message is about Kazakhstan as a place of peace, unity, and multiculturalism. Authorities are trying to be diplomatic—perhaps too much so.

As for the media—by law, it operates in two languages, and 50% of content must be in Kazakh. All TV and radio stations must have content in both Kazakh and Russian. Some fully Kazakh-language platforms exist, and Russian-language ones also function freely, including broadcasts from Russia. In fact, Russian television and online content are widely accessible and popular. This creates a complex media environment where narratives vary widely, but people usually choose based on comfort and habit. There are many voices calling to shut down free broadcasting of Russian TV because of war propaganda and racist claims. Some Russian TV channels were cancelled in 2024, but many still broadcast, as the responsible minister claims, "It is a market. Demand moves supply" (Zhulmukhametova, 2024).

In sports and tournaments, Kazakhs, Russians, and athletes of other ethnic origins participate together. Teams in high schools or universities are often mixed. In professional sports like football (soccer), both ethnicities are represented. Some famous players are ethnic Russians, some Kazakhs, and it is not even a topic of debate. Broadcasting of sports games alternates between Kazakh and Russian commentary.

In cities like Almaty or Astana, people of different backgrounds walk in the same parks, shop in the same malls, and attend the same concerts and festivals. Public places like libraries, skating rinks, or book fairs are for everyone (Fig. 6). Digital spaces like gaming cafés or co-working hubs are also ethnically neutral (Fig. 7).

Historically, some Germans left for Germany in the 1990s but still around 200,000 remain in Kazakhstan. Jews went to Israel. Koreans stayed. Some are now trying to revive their culture and language. There are complaints about private companies hiring based on ethnicity, but such discrimination is illegal. Officially, the Constitution grants equal opportunity.

English-language education is growing fast. In schools, it is taught from early grades, and many universities now offer full programs in English. Students prefer to know three languages: Kazakh, Russian, and English. That is the norm for the educated urban youth. Teaching of history has definitely changed since 1991. Soviet versions of history are no longer accepted. However, Russian government efforts to have joint history textbooks are still under discussion. The focus now is on Kazakh statehood, resistance to colonization, and local heroes. But rewriting history is complicated. Some textbooks are better than others. The main narratives are written by historians working under government guidance, often blending patriotism with diplomacy.

In terms of common ground it is not about formal politics, but about shared spaces. People mix at festivals, sporting events, in shopping centers, on Instagram, and in cafés. That is where identity becomes fluid, relaxed, and shared. But tensions

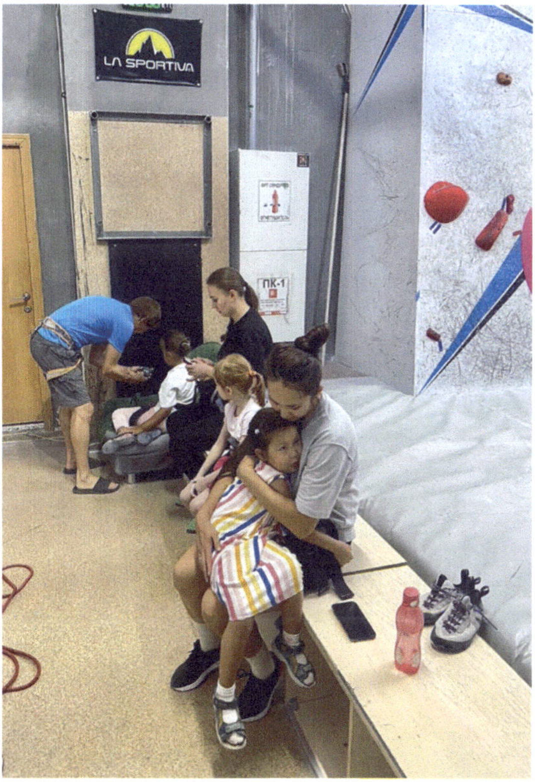

Fig. 6 Russian and Kazakh people are in the rock-climbing gym in Almaty. Mothers and fathers are safety rope partners for the kids. On the back we can see the fire extinguishers instruction in two languages. (Photo by the author, August 11, 2025)

can rise when language, memory, or war are politicized. Still, daily life is more about co-existence than conflict. So far, it holds.

Conclusion: Kazakh Russians and Russian Kazakhs

The world of Kazakhs and Russians in Kazakhstan are parallel, co-existing and interacting every day in history. Questions on past violence, domination and historical grievances are in the process of resolution as national building takes place in Kazakhstan. There are other ethnic groups in Kazakhstan along with the Kazakhs and Russians and their rights, freedoms and the future are important for peace and economic well-being of the country. The road to national unity, economic development, and civil peace is not straight and requires daily conscious effort from the state and people.

Despite the strong gravitational influence from Russia towards Kazakhstan and ethnic Russians, we should acknowledge that differences exist between Russians who live in Kazakhstan and Russians who live in Russia. The differences were obvious when many were running from Russia and mobilization was declared in 2022.

Fig. 7 Russian and Kazakh men selling mobile devices in the central mall (TSUM) of Almaty. (Photo by the author, August 11, 2025)

The familiarity with Kazakh everyday life, food, power and legal structures, family and regional features as well as lack of implicit imperial feelings were front and center in discussions. These are the differences between some ethnic Russian locals and Russians living in Russia. These cultural differences are not significant enough to suggest the emergence of a distinct ethnic or political group such as “‘Kazakh Russians” or “Kazakhs of Russian origin”. Instead, the unifying compromise is the shared identity of being Kazakhstani, common to all citizens of the republic. This civic and ethnonationalist move creates a certain ambiguity. As Ó Beacháin and Kevlihan (2013) correctly note “strategic ambiguity between civic and ethnonationalist tendencies has played an important role in political and social developments since independence….” From another side and after 20 years since this publication and the Russian invasion to Ukraine, there is a case of a certain questions and loyalty in identity construction in Kazakhstan. It is an ongoing process with constant effort.

There is not any specific new emerging special ethnic group of Kazakh Russians or Kazakhs of Russian origin. Mastery of the Kazakh language is still weak among many. The domination of Russia in information and entertainment is strong but Russians who have lived in Kazakhstan for several generations and who decided to connect their future with Kazakhstan might have their own specific traits as they were merging in different political and cultural contexts. Russified Kazakhs also are some kind of a special phenomenon as many do not speak Kazakh well and are forming their own social group, which is not Kazakh or Russian, but as a group that exists between the two worlds. They are a bridge between the two groups, but also a group which is at the crossroads. These two worlds of the Kazakhs and Russians are not confrontational but have certain tensions on politics, past privileges and language as a marker of identity and respect within Kazakhstan. Both groups have their own dynamics in the development of social, demographic, and economic processes that influence their interactions and visions for the future.

References

Agency of the Republic of Kazakhstan for Strategic Planning and Reforms, National Bureau of Statistics. (2023). *National composition, religion and language proficiency in the Republic of Kazakhstan: Results of the National Population Census of 2021 in the Republic of Kazakhstan.* National Bureau of Statistics.

Akorda.kz. (1995). Constitution of the Republic of Kazakhstan. Official Website of the President of the Republic of Kazakhstan. https://www.akorda.kz/en/constitution-of-the-republic-of-kazakhstan-50912

Äliakbar, Ä.-S. (2019, November 6). The ancient cities of Kazakhstan and Siberia were not built by Russians [Qazaqstan men Sibirdiñ köne qalalaryn orystar salğan joq]. *Abai.kz.* https://abai.kz/post/100334

Almas, S. (2025, August 2). Support grows for establishing an Orthodox Church independent from Moscow in Kazakhstan *[Za otkrytie nezavisimoy ot Moskvy pravoslavnoy Tserkvi vystupili v Kazakhstane]. Exclusive.* https://exclusive.kz/za-otkrytie-nezavisimoj-ot-moskvy-pravoslavnoj-cerkvi-vystupili-v-kazahstane/

Auyezbek, A. (2023, April 17). How war affects immigration of Russians from Kazakhstan to Russia. *CABAR.asia.* https://cabar.asia/en/how-war-affects-immigration-of-russian-from-kazakhstan-to-russia

Azar, I. (2014, October 20). Ust-Kamenogorsk People's Republic: Are Russians in Kazakhstan waiting for the "polite people"? *Meduza.* https://meduza.io/news/2014/10/20/ust-kamenogorskaya-narodnaya-respublika

Bissenova, A. (2019). Social change unsettles Kazakhstan. *Current History, 118*(810), 271–275. https://www.jstor.org/stable/48614467

Bolatuly, T. (2024, June 25). *Lenin was removed, Ridder was not installed: Debates about the empty pedestal in East Kazakhstan* [Ленина снесли, Риддера не установили. Споры вокруг пустующего постамента в Восточном Казахстане]. Azattyq. https://rus.azattyq.org/a/east-kazakhstan-region-ridder-monument/33007956.html

Buttino, M. (2003). *La rivoluzione capovolta: L'Asia centrale tra il crollo dell'impero zarista e la formazione dell'URSS.* l'Ancora del Mediterraneo.

Cameron, S. (2018). *The hungry steppe: Famine, violence, and the making of Soviet Kazakhstan.* Cornell University Press.

Census. (1974). Results of the All-Union Population Census of 1970 [*Itogi Vsesoyuznoy perepisi naseleniya 1970 goda*]. Vol. 7, p. 117. Statistika.

Chulanova, Z. K. (2021). External migration of the population of Kazakhstan during the years of independence [Внешняя миграция населения Казахстана за годы независимости]. *The Scientific Heritage, 76,* 37–45. https://cyberleninka.ru/article/n/vneshnyaya-migratsiya-naseleniya-kazahstana-za-gody-nezavisimosti

Darimbet, N. (2017, January 21). Why are Kazakhs the poorest in Kazakhstan? [Pochemu kazakhi bednee vsekh v Kazakhstane]. *Ratel.kz*. https://ratel.kz/raw/pochemu_kazahi_bednee_vseh_v_kazahstane?page=3

Dave, B. (2007). *Kazakhstan: Ethnicity, language and power*. Routledge.

Demoscope. (2025, February 20). *Poll: The war in Ukraine continues to polarize Kazakh society* [Опрос: Война в Украине продолжает поляризовать общество Казахстана]. https://demos.kz/opros-vojna-v-ukraine-prodolzhaet-poljarizovat-obshhestvo-kazahstana/

Dugin, A. (1997). *The foundations of geopolitics [Osnovy geopolitiki]*. Arktogeja.

Duvanov, S. (2023, May 3). Interview by M. Sugirbayeva: Why non-Kazakhs in Kazakhstan have a "tenant syndrome." *Exclusive.kz*. Retrieved July 15, 2025, from https://exclusive.kz/pochemu-u-kazahstanskih-ne-kazahov-sindrom-kvartirantov/

Exclusive.kz. (2025, May 7). *It became known who was behind the calls to participate in the "Immortal Regiment" rally in Kazakhstan for monetary reward* [Stalo izvestno, kto stoyal za prizyvami k uchastiyu v aktsii "Bessmertnyy polk" v Kazakhstane za denezhnoye voznagrazhdenie]. https://exclusive.kz/stalo-izvestno-kto-stoyal-za-prizyvami-k-uchastiyu-v-akczii-bessmertnyj-polk-v-kazahstane-za-denezhnoe-voznagrazhdenie/

Gall, C. (2022, April 11). Bucha's month of terror. *The New York Times*. https://www.nytimes.com/2022/04/11/world/europe/bucha-ukraine-russia-war.html

Goble, P. (2021, December 2). Violence spreads in Kazakhstan, and now involving attacks on ethnic Russians. *Eurasia Daily Monitor, 18*(180). The Jamestown Foundation. https://jamestown.org/program/violence-spreads-in-kazakhstan-and-now-involving-attacks-on-ethnic-russians/

Hirsch, F. (2005). *Empire of nations: Ethnographic knowledge and the making of the Soviet Union* (1st ed.). Cornell University Press. http://www.jstor.org/stable/10.7591/j.ctt1tm7j58

Hochu_zhyt. (2024). [Telegram post]. Telegram. https://t.me/hochu_zhyt/3220

Ibadildin, N. (2024). Kazakh nomads: Thousand-year routes, environment and political oppression. In S. D. Brunn (Ed.), *Geography of time, place, movement and networks: Volume 2: Mapping heritage journeys and sameness (Chapter 8)*. Springer. https://doi.org/10.1007/978-3-031-58029-1_8

Karabek, E. (2011, September 7). The Uralsk events of 1991: The beginning of the conflict [Uralskie sobytiya 1991 goda: Nachalo konflikta]. Azattyq Radiosy. https://rus.azattyq.org/a/oral_okigasy_uralskie_sobytia_uralsk_events_1991/24320134.html

Kendirbai, G. (2002). The Kazakh intelligentsia and the Alash Orda movement. *Journal of Central Asian Studies, 7*(1), 1–18.

Khalid, A. (2021). *Central Asia: A new history from the imperial conquests to the present*. Princeton University Press. https://doi.org/10.2307/j.ctv19qmf3k

Lillis, J. (2014, April 14). Kazakhstan: Russians blend loyalty to Nazarbayev with pro-Kremlin sentiments. *Eurasianet*. https://eurasianet.org/kazakhstan-russians-blend-loyalty-to-nazarbayev-with-pro-kremlin-sentiments

Martin, V. (2001). *Law and custom in the steppe: The Kazakhs of the Middle Horde and Russian colonialism in the nineteenth century*. Routledge.

Morrison, A. (2021, July 1). Russia is allergic to the word colonialism: The 1916 Uprising through politics and history]. *Радио Азаттык*. https://rus.azattyq.org/a/u-rossii-allergiya-na-slovo-kolonializm-vosstanie-1916-goda-cherez-prizmu-politiki-i-istorii/31334649.html

Nurmukhanbetov, M. (2022, June 13). "We built cities for you…" ["My vam goroda ponastroili…"]. *Check-Point.kz*. https://check-point.kz/publication?id=557

Ó Beacháin, D., & Kevlihan, R. (2013). Threading a needle: Kazakhstan between civic and ethno-nationalist state-building. *Nations and Nationalism, 19*(2), 337–356. https://doi.org/10.1111/nana.12022

Olcott, M. B. (1993). Kazakhstan: A republic of minorities. In I. Bremmer & R. Taras (Eds.), *Nation and politics in the Soviet successor states* (pp. 340–376). Cambridge University Press.

Oshuev, N. (2025, July 31). German Archive: Putin claimed Crimea and Kazakhstan as early as 1994 [Архив ФРГ: Путин в 1994 году претендовал на Крым и Казахстан]. *Deutsche Welle*.

Payne, M. J. (2011). The "Forge" of the Kazakh proletariat? The Turksib, nativization, and industrialization during Stalin's First Five-Year Plan. *Slavic Review, 70*(4), 746–771.

Pianciola, N. (2019a, October 30). *Nomads and the State in Soviet Kazakhstan*. Oxford Research Encyclopedia of Asian History. Retrieved July 17, 2025, from https://oxfordre.com/asianhistory/view/10.1093/acrefore/9780190277727.001.0001/acrefore-9780190277727-e-314

Pianciola, N. (2019b). Scales of violence: The 1916 Central Asian uprising in the context of wars and revolutions (1914–1923). In A. Morrison, R. Drieu, & A. Chokobaeva (Eds.), *The Central Asian revolt of 1916: A collapsing empire in the age of war and revolution* (pp. xx–xx). Manchester University Press. https://doi.org/10.7765/9781526129437

Polian, P. (2004). *Against their will: The history and geography of forced migrations in the USSR*. Central European University Press.

Ranking.kz. (2025, June 24). *The grass is greener where we are: Emigration from Kazakhstan hits a record low*. https://ranking.kz/en/reviews-en/society/the-grass-is-greener-where-we-are-emigration-from-kazakhstan-hits-a-record-low.html

Romanov, A. N. (Comp.), & Shchetinin, M. P. (Eds.). (2008). *This is the pride of the Altai Krai: Based on materials from a creative competition* [Этим гордится Алтайский край: по материалам творческого конкурса].

Tarenchikova, A. (2025, April 15). Interethnic marriages in Kazakhstan: How many are there [Mezhetnicheskie braki v Kazakhstane: skol'ko ikh]. *Inbusiness.kz*. https://inbusiness.kz/ru/last/mezhetnicheskie-braki-v-kazahstane-skolko-ih

Verkhovny Sovet of the Kazakh SSR. (1990, October 25). *On the Declaration of State Sovereignty of the Kazakh Soviet Socialist Republic* [O Deklaratsii o gosudarstvennom suverenitete Kazakhskoy Sovetskoy Sotsialisticheskoy Respubliki], No. 307-XII. https://adilet.zan.kz/rus/docs/B900001700_

Verkhovny Sovet of the Republic of Kazakhstan. (1991, December 16). *On the state independence of the Republic of Kazakhstan: Constitutional law of the Republic of Kazakhstan*. https://online.zakon.kz/Document/?doc_id=1000798&pos=4;-106#pos=4;-106

Verkhovnyy Sovet Respubliki Kazakhstan. (1993, January 28). Constitution of the Republic of Kazakhstan (No longer in force). [*Konstitutsiya Respubliki Kazakhstan*]. https://adilet.zan.kz/rus/docs/K930001000_

Vishnevsky, A. G., & Shkolnikov, V. M. (1997). *Mortality in Russia: Major risk groups and policy priorities*. Carnegie Moscow Center.

Zatulin, K. F., & Migranyan, A. M. (1997, March 26). CIS: the beginning or the end of history. Towards a change of milestones. *Nezavisimaya Gazeta* [SNG: nachalo ili konets istorii. K smene vekh. Nezavisimaya gazeta. Sotrudnichestvo].

Seeking Togetherness at the Table: Missionaries Wrestling with Everyday Encounters with Parallel Worlds in Contemporary Capitalist Society in Cajamarca, Peru

Iain Ross

> If worlds are multiple, then the possible must also be multiple. […] another world is possible because another real and another possible are possible. That other world is a world where many worlds fit, or the pluriverse. (Escobar, 2020: ix)

> As we imagine and work toward another possible world, it is important to reemphasize that mestizaje ["mixedness"] is not an abstract romantic category. […] Beyond a simple coming together of elements, […] mestizaje corresponds to the day-to-day incarnated reality that people confront where their faith plays a crucial role in challenging inherited worldviews and in providing a sense of coherence and hope. (Medina, 2021: 53)

> Eating with the fullest pleasure […] is perhaps the profoundest enactment of our connection with the world. In this pleasure we experience our dependence and our gratitude, for we are living in a mystery, from creatures we did not make and powers we cannot comprehend. (Berry, 2010: 152)

Introduction

Radical Pluralism in the Anthropocene

In the light of the ecological degradation and injustices posed by climate change, the discussion of a new geological epoch known as the Anthropocene, and the effects of techno-capitalist expansionism, radical onto-epistemological pluralism has presented itself in recent years as an avenue towards better futures in the social sciences and humanities. The renown Columbian-American anthropologist Arturo Escobar's

I. Ross (✉)
Cajamarca, Peru

Dundas, ON, Canada
e-mail: iain.ross@alumni.utoronto.ca

S. D. Brunn (ed.), *Transdisciplinary Research Across Knowledge Worlds*,
https://doi.org/10.1007/978-3-032-19491-6_7

(2018) concepts of the "One World World" and the "pluriverse" are representative of this pluralist/anti-universalist movement in contemporary academia.

Escobar's project rejects what he names the "One World World," the universalist and universalising character of hegemonic techno-scientific capitalism and Euro-American imperialism which seek the assimilation of every area of life of all people(s) from their most basic epistemological and ontological categories all the way up to their cosmology and everything in between. He describes the One World World like this:

> The notion of oww [One World World] signals the predominant idea in the West that we all live within a single world, made up of one underlying reality (one nature) and many cultures. This imperialistic notion supposes the West's ability to arrogate for itself the right to be 'the world,' and to subject all other worlds to its rules, to diminish them to secondary status or to nonexistence, often figuratively and materially. (Escobar, 2018: 86)

As an alternative, Escobar works toward a "pluriverse," a situation in which radically different, other-than-capitalist subjectivities and socio-environmental assemblages can coexist. This project is progressed through the recognising, validating, and empowering of non-hegemonic "world-making" practices, especially among indigenous and afro-descended peoples. For Escobar, to fight for justice is to fight for the recognition and protection of the other worlds that non-hegemonic peoples inhabit.

The pluriverse goes beyond decentring Euro-American hegemony; it has no centre. In the same way, it works towards collapsing metaphysical-dualisms deeply established in the history of Western thought, specifically mind/body, nature/culture, subject/object and thinking/feeling. Other prominent academics working in this academic current include American political theorist Jane Bennett (2009), Chinese American anthropologist Anna Tsing (2015), Peruvian anthropologist Marison de la Cadena (2015), Indian geographer Neera Singh (2018) and Spanish interdisciplinary academic Maria Puig de la Bellacasa (2017). They encompass several more specific theoretical streams including New Materialism, New Animism, affective ecologies and Anthropocene studies.

For many in the social sciences, addressing the present climate crisis requires a rejection of anthropocentrism, which emerges from the dualistic, non-relational ontologies and epistemologies listed above. Over the course of the modern era, the conjunction of Western Christian theology, European colonialism and the emergence of capitalism have led to a predominantly anthropocentric experience of the world in hegemonic Western culture (Jennings, 2019). Escobar's pluriverse project, and the radical pluralist/anti-universalist current generally, attempts to support the creation of space for non-anthropocentric life-ways that retain the sense of interconnectedness that secular capitalism has denied many modern Western people with disastrous environmental consequences.

The Christian Creator

Theology in any major branch of orthodox Christianity (i.e. Protestant, Catholic and Orthodox) is unlikely to concede that there exist multiple worlds in the strict ontological sense Escobar describes. The reason is that some core theological convictions of orthodox Christianity depend on the idea that *everything* that exists was created by one God (Fergusson, 2011). For instance, the idea that Christians can be confident of God's ultimate victory over all the powers of darkness has traditionally rested on the idea that God is the creator of all. The concern is that if there existed some thing that God had not created, that thing could conceivably be more powerful than God, or at least capable of resisting God indefinitely. This millennia-old discussion is the reason why the ontology of evil (and *Satan*, evil's personification) has traditionally been described as a kind of *lack* or *absence* of good rather than something with an independent existence of its own. This idea can be traced to Augustine of Hippo, a fourth century North African theologian who is a towering authority in the Christian theological tradition. In this way, Christian theology affirms that all that truly exists was created good, by one good Creator.

With this in mind, any orthodox Christian discourse of multiple or parallel "worlds" would always need to be couched in the idea that these worlds ultimately belong to one wider cosmos or universe of which God is the only creator. This idea is challenged when Escobar (2018: 86) writes that it is the imperialist and Western One World World that understands the world to be "*made up of one underlying reality (one nature) and many cultures*." The idea that the God of Christianity (and Judaism and Islam for that matter) is the creator of all that exists has its deepest roots not in European imperialism, however, but in the in the Near East and North Africa, predating by at least a millennium modern capitalism or Euro-American cultural hegemony.

"The World" in Christianity

This notion that a significant amount of Christian theology hangs on the idea of there being only one Creator of one universe has not, however, precluded complex and nuanced experiences and theologies of parallelism within Christianity. In fact, the opposite is true; the parallel yet overlapping existence of the temporal and the eternal, of heaven and earth, of God's space and humanity's space is equally central to Christian thought and experience as a single Creator. An accessible window for those unfamiliar with Christian theology into these parallelisms can be found in the word "world" which, in English translations of the Bible, has a variety of paradoxical uses. Here are a few examples [emphasis added]:

> For God so loved the *world*, that he gave his only Son, that whoever believes in him should not perish but have eternal life. John 3:16

> You adulterous people! Do you not know that friendship with the *world* is enmity with God? Therefore, whoever wishes to be a friend of the *world* makes himself an enemy of God. James 4:4

> He said to them, "You are from below; I am from above. You are of this *world*; I am not of this *world*." John 8:23

> For everyone who has been born of God overcomes the *world*. And this is the victory that has overcome the *world*—our faith. 1 John 5:4

> Religion that is pure and undefiled before God the Father is this: to visit orphans and widows in their affliction, and to keep oneself unstained from the *world*. James 1:27

Any attempt to unify these texts in a coherent way must wrestle with the question of whether the "world" should be loved or hated; what part of God or the Christian is "not of the world" and *where* that part exists; and why the world is something to be "overcome" and something that can "stain".

Christian thought generally attempts to make sense of this by understanding the "world" or "worldly things" to refer to the evil or sinful social structures, values and practices that appear *inherent* to the human condition and putting these in opposition to "the Kingdom of God". The Kingdom of God is a transcendent realm that exists within but also beyond the "world" or as a result of the world's transformation or purification. (I use multiple similar terms here to represent the widest possible set of Christian theological streams, which nuance these concepts in specific ways in accordance with their wider theological framework). The Kingdom of God is the realm in which all that which opposes God's loving rule is no longer present because it has been renewed or redeemed or re-created or removed. The arrival of this Kingdom is an ongoing process leaving the Christian (or perhaps all Creation) in a liminal space of ongoing renewal, redemption, or re-creation. Every practicing Christian reckons with the idea that in some way they partially or incompletely inhabit the Kingdom of God, while also continuing to inhabit "the world." In some Christian communities, this complex idea is summed up in the idea of being "*in* the world but not *of* the world."

What this thinking means in practice that can vary greatly from individual to individual, from congregation to congregation and from denomination to denomination. It can, for example, imply abstaining from certain cultural practices (like listening to a particular genre of music or participating in holiday celebrations like Halloween); abstaining from participating in certain professions (like being a soldier in the military or a police officer); abstaining from any kind of violence (i.e. pacifism as a religious doctrine) or participation in government (including voting or hold any government office); or abstaining from certain practices deemed "worldly" (like gambling, drinking alcohol, cheating, bribery or sexual promiscuity). It can also imply establishing geographically separate communities (something that the Amish and Mennonites are known for). Christian theology and Christian congregations around the world have been continually wrestling with these parallel concepts of God's love for Creation, the Kingdom of God and "the world" as moral decay for nearly two millennia. Grappling with this tension is a focus of many Christians' spiritual formation over the course of their lives, both as individuals and in community.

Christianity and the One World World

This complexity and the long, diverse and sometimes disparate set of traditions within Christianity are not sufficiently addressed by the social theorists mentioned above, especially in the areas of New Materialism, New Animism, Anthropocene studies and affective ecologies. In fact, ironically, there is occasionally a tendency in radical social and political theoretical writing to reify a kind of dualism (of precisely the kind that Escobar rejects) between the liberationist radical and the imperialist/capitalist mainstream, between the oppressor and the oppressed, the West and "the rest", Western Christianity and indigenous religion. The history and global present of Christianity do not fit neatly into any of these dualisms. Christian theology, and Christian congregations around the world have also been wrestling with the colonialist, capitalist, and arrogantly anthropocentric influences that Escobar opposes within in their faith for centuries.

The complicity of many Christian churches, and Christian theologies, in heinous and sustained structures of violence and injustice of various kinds is undeniable (Jennings, 2011; Keller, 2021; Sugirtharajah, 2022). Characterising this complicity as representative of Christianity as a whole, however, would involve the erasure of the long, fraught, and storied traditions described above as separation from "the world," which for some anti-establishment Christian traditions in the West (like the Anabaptists) has implied strict pacifism and non-participation in government. It would also deny the dizzying diversity of Christian thought and experience across the world, not least the varieties of Christian life, practice and thought that originate in the Global South. (Liberation theology is one key example of this, Integral Mission is another. Ethiopian Orthodoxy is yet another having an unbroken history since the fourth century CE and never having been colonised by a European power). Christian communities exist in almost every kind of social, cultural, economic, political, ethnic and environmental assemblage imaginable. How Christianity is expressed has changed through time and space for nearly 2000 years across almost every region of Earth.

Christian agrarianism (Berry, 2002; Brueggemann 1991, 2000; Davis, 2009; Wirzba, 2007), for example, is re-discovering place-based belonging, affective relational ontologies and agro-ecology within the Christian Scriptures and tradition. The discussion of the possibility of a Christian New Animism has been ignited by Mark Wallace (2020) in *When God was a Bird*. Also, an increasing number of academic establishments now offer courses or entire degrees in "World Christianities" and "contextual theology". An example of contextual theology that is relevant for this chapter is the way theologian Néstor Medina has re-appropriated the Spanish racial term *mestizaje* (mixedness) as a concept representing the "resilient, persistent, even obstinate lived faith [...] of Latina/o communities." (Medina, 2021: 48).

It is in the incomplete, ongoing process of mixing worlds in our embodied everyday life that lived faith is evident. All of this points to the decolonisation, indigenisation, and shifting-centre of Christian thought and practice away from Europe that has been taking place for at least a century.

By addressing Christianity, I do not exclude other religions. On the contrary, in my experiences in academia, the rich, complex and diverse histories of thought of other world religions are also inadequately addressed in social and political theory. However, I only have the knowledge and experience to discuss Christianity. There is a need for create space for theologians and religious scholars from within other religions to contribute to the issues raised by the radically pluralist social and ecological theorists in New Materialism, New Animism, Anthropocene studies and affective ecological studies.

Encounter and Togetherness

In summary, within contemporary and historical Christianity there exists both a universalist affirmation of the unity of all Creation under one Creator and simultaneously multiple parallelisms: including, but not limited to, that of "the kingdom of God" and "the world" and the coloniser and colonised. What this thinking can bring to the table in social theory is to suggest a redirection toward *encounter* and *togetherness* and away from radical metaphysical pluralism. To describe reality as series of worlds with no underlying connection between them would limit or even eliminate the possibility of encounter and togetherness. Togetherness does not imply assimilation. In fact, it requires difference. Togetherness is peaceful and just sharing of time and place. The ongoing process, quality, and relational outcome of encounters with the Other (including non-human Others) in postcolonial capitalist contexts is the conceptual centre of this chapter. Its key questions are: where and how do parallel worlds encounter one another? What fosters togetherness and what are the obstacles to togetherness when worlds collide?

Parallelism and the Contemporary Christian Missionary in Cajamarca

Methodology and Definitions

This chapter explores these parallelisms in the thought and embodied experience of nine contemporary Christian missionaries in Cajamarca, Peru. Cajamarca is an agricultural region in the northern Peruvian Andes known for its dairy production and its extravagant Carnival celebrations in February or March. It is also known as the place where the last Incan king, Atahualpa, was captured and killed by Pizarro, the

Spanish Conquistador. Figure 1 shows an informative plaque aimed at tourists in the main square of Cajamarca in the historic centre. The title reads "Cajamarca: the city of the encounter of two worlds" (author's translation). The other plaques around the historic centre elaborate on this idea of encounter, describing and depicting graphically these historic events in Cajamarca as the beginning of a deep mixing and melding of native and European cultures.

This can be seen as example of Medina's use of the idea of *mestizaje* or "mixedness" mentioned above; it reflects a history and contemporary realty of a constant mixing of worlds, not just a racial or ethnic mixing. Cajamarca is profoundly *mestizo* (mixed) in the broadest possible sense. Part of the complex history of identity politics in Peru means that it is very uncommon for people in Cajamarca to identify as indigenous. People have differing types and degrees of connection to the ways of their ancestors. However, unlike in neighbouring Ecuador and Bolivia, in Peru even the communities most connected to traditional Andean culture are unlikely to describe themselves as indigenous. Identity politics has been dominated by the language of class (peasant vs. landowner) and geography ("coastal", "mountain" and "rainforest" people) in Peru (Andolina et al., 2009; Greene, 2006). Peruvian anthropologist de la Cadena has argued that discourses of *mestizaje* (mixedness) constitute a kind of erasure or white-washing of indigenous Peruvians (de la Cadena, 1998). Indeed, the politically salient options for self-identification among traditional (and

Fig. 1 Tourist information plaque in the main square of the historic centre of the city of Cajamarca (Source: author)

especially rural) mountain Peruvians are limited by the weakness of discourses of indigeneity in Peru. However, in line with Medina's understanding of *mestizaje*, the very widespread *mestizo* identity in Peru is itself a political, social, and even spiritual identity that, though fraught, creates as many political opportunities as it erases.

Nine individuals (including myself) participated in this study through semi-structured interviews (or keeping a journal in my case). The term "missionary" itself is contested within the diverse expressions of Christianity. Some groups remain comfortable with the term, others prefer to adapt it to mitigate the term's colonial baggage ("missionary partner" for example), still others avoid the term altogether. Additionally, while the idea of Christian mission generally implies the notion of going somewhere *distant*, it is not uncommon to understand this in terms of *social, economic, cultural* or *religious* distance, not geographical distance. Therefore, while to some mission still denotes travelling to another part of the world to engage in the propagation of Christianity, to others engaging in Christian mission may equally refer to sharing meal with people experiencing homelessness in their own neighbourhood.

The term "evangelical missionary" is used in this discussion to refer to the participants. That definition does not necessarily imply, however, that all participants would comfortably describe *themselves* as a "missionary" nor "evangelical." Each participant had a different relationship and understanding of these terms and the extent to which they identify with the words. When asked how they understand the phrase "Christian mission" and their involvement in it, there was notable diversity even within such a small set. Several participants resisted the idea of reducing Christian mission to proselytism, or basic assistance programmes like community soup kitchens. One participant, for example, works in social-justice public policy advocacy across Latin America and understands this effort as part of the mission of the Church. As such, the term *evangelical missionary* should be read here as a heuristic device to refer to people who have travelled to Cajamarca from another country and engage in explicitly Christian activities in congregations commonly referred to as *evangélica*, even if "Christian mission" in the traditional sense of international proselytism is not what brought them to Cajamarca or kept them here.

Four of the participants are women, five are men; the group included citizens of the United States, Brazil, the United Kingdom, and Canada. Some participants had been in Cajamarca only a few days, others for several months or years. The participants belonged to a variety of Protestant denominations (including Assemblies of God, Calvary Chapel, Be in Christ, Non-denominational independent congregations, and others). In Cajamarca, these congregations would all generally be described as *evangélica* [evangelical]. About half understand their reason for being in Cajamarca as primarily to be missionaries and were in regular communication with an organization that sent them. the other half were in Cajamarca for broader or more personal reasons but engaged, nonetheless, in explicitly Christian activities (teaching the Bible to children in a church, for example). Eight lived or stayed in the city of Cajamarca and were not separated spatially from the rest of society in Cajamarca. That is, they frequented the same shops, pharmacies, hospitals, restaurants and markets as people in Cajamarca and their neighbours were people from

Cajamarca. They communicated daily with people from Cajamarca in person or on their phones (generally through WhatsApp, which is by far the most popular messaging application in Peru). One missionary was based in a private bible college in the agricultural outskirts of the region of Cajamarca and thus had a certain level of geographical separation. However, he pastored a church in a neighbouring town whose congregants were primarily local residents and so also had regular in-person contact with people from Cajamarca. Even this very small sample of participants displayed diverse political and theological positionings on a range of issues, demonstrating again the diversity of global evangelicalism.

Justification

Christian missionaries constantly grapple with the multiple tensions described above in their daily lives, often in quite high-stakes circumstances. Being "in the world but not of the world" may imply something quite different in Cajamarca than in their community of origin. Acceptance, belonging and relationship building in Cajamarca may depend on rapidly adapting to a different understanding of the concept. Formulating and reformulating a working understanding of what is universal about the human condition versus what is particular to their culture of origin or the culture of Cajamarca is a seemingly never-ending and sometimes exhausting task. They must also grapple with the tension between the inescapably "universalising" desire to be a mouthpiece for God's desire to redeem all people(s) everywhere (indeed, all creation), and the ever-present danger of instead universalising the violence, pride and selfishness that dwells in the heart of the missionary, or in the wider society of origin to which the missionary willingly or unwillingly belongs. The everyday, voluntary, and embodied way that these parallelisms and tensions play out in the lives of contemporary missionaries make them an interesting and important group through which to critically engage with Escobar's One World World and the associated theoretical currents in contemporary social science.

This chapter is not an attempt to speak normatively on any of the issues described nor to make any theological or political claims. It also does not claim to be representative of any other missionaries. The objective is to reveal not conclusions, but possibilities and processes involved in "world-making" or "world-negotiating" in the context of postcolonial capitalist encounter. The observations have a geographical flavour. That is, they pay particular attention to the spatial and embodied experience.

Results

A Patchwork of Overlapping Parallelisms

There are many layers of parallelism which overlap in a complex patchwork in the everyday lives of the missionaries. Three areas of parallel experience emerged as significant. Firstly, missionaries come to Cajamarca in the context of preexisting ideas (held by both missionaries themselves and residents) about race and the relationship between Peru and their country of origin. Secondly, missionaries come with a set of moral convictions about what being "in the world but not of the world" means to them in practice. Christian congregations in Cajamarca also have a set of moral convictions. These moral imaginations can overlap at times and cause friction at others. Missionaries must learn to navigate the sometimes-unfamiliar moral landscapes. Thirdly, the dynamics and geographies of *class* that missionaries encounter in Cajamarca can be quite different from those in their country of origin. More specifically, the missionaries' social location and their ability (and desire) to interpret, navigate, leverage and/or resist class dynamics may be markedly different than it is or was in their country of origin.

One place emerged consistently across every interview as a locus of encounter and a focal point in the struggle for togetherness: the shared table. The shared table was described as a universal human practice, a Latin American cultural practice and a Christian spiritual practice. It was described as a place where relational tension, social distance and moral distance, existed but also where belonging, transformation and justice were revealed and experienced.

National and Racial Stereotypes

Preexisting ideas about the relationship between Peru and their country of origin can foster either unity or separation between missionaries and the people they serve. Arthur, a Brazilian missionary only briefly in Cajamarca to promote participation in a social-justice orientated online training course, immediately and animatedly described the shared colonial legacy of Latin America as his starting point for identification with people from Cajamarca:

> Before anything else, I am Latin American. So by some means we are siblings of one sole homeland. [...] Peruvians and Brazilians, we are siblings of the same Great Homeland as some call [Latin America]. We have very similar challenges because we are the result of very similar processes of colonization. [translation by the author]

On the other hand, Carys, a British woman who lived in Cajamarca for several months over several years, spoke of the stereotypes associated with white people and the barriers that they created to connection, particularly with adults.

> When adults see a white person in Cajamarca, it comes with a lot of stereotypes.[…] When adults saw a white person, it didn't feel like we were starting off a relationship on an equal footing it felt like they presumed that white people were better, but there wasn't that with children, it felt a lot more equal.

Separation from "The World"

The idea of being "in the world but not of the world," exists both in Cajamarca and in the countries of origin of missionaries, but it can be expressed differently. Carnival, for instance, (see Fig. 2) is an event that many evangelicals in Cajamarca consider too "worldly" for participation to be permissible because of the excessive alcohol consumption and sexual licentiousness that is commonplace during the festivities. For British missionaries, while excessive alcohol consumption and licentiousness are hardly absent in British culture, Carnival can be a wholly foreign experience with no familiar reference point. This may render British missionaries' working understanding of how to remain "unstained" by the world unable to help them decide on an appropriate level of participation in Carnival in Cajamarca as a cultural event, at least at first.

Another example are the expectations surrounding gender roles. Carola was born in Cajamarca but spent much of her adult life in the US and has now returned to Cajamarca as a long-term missionary. Aron, her husband, is a Latino US citizen. They described having to wrestle against restrictive ideas about gender in Cajamarca, having experienced more freedom in the US. They understand this to be a primarily cultural issue:

Fig. 2 Revelers partying in the streets of Cajamarca on the central day of Carnival celebrations (Source: author)

> Carola: To me, language and culture are not a hindrance. I'm used to the culture, my family is from here. […] To me the only thing that really makes me feel that I need to fight…is that I am a woman in ministry.
> Aron: …that doesn't always wear a skirt.
> Carola: …and I don't conform to the patterns; I refuse to do that. Cause I think that is just culture it's not God. I didn't wear skirts in Kentucky and I feel like I'm being fake if I just conform. I don't do that.

However, I am aware that for some conservative evangelicals from Cajamarca, the expectation that women should not be in church leadership and that they should wear long skirts and not trousers are not connected to so much to culture but to the idea of being separate from things they consider "worldly" like feminism or female overambition.

Despite these parallel moral landscapes, the encounter of evangelicals from Cajamarca and foreign missionaries creates opportunities for (re)negotiating what this principle should mean for both parties through biblical or theological (re)interpretation and (re)application. The specific content of what the missionaries thought they should remain separate from will be discussed below. What I wish to highlight at this point is that in addition to the parallelism inherent in the idea of being "in the world but not of the world," missionaries face another tier of parallelism through the potential for significant differences in their understanding of the idea compared to those of the people they are working with. In this situation, missionaries can adapt their understanding or lifestyle choices to align with the expectations of evangelicals from Cajamarca; they can attempt to influence evangelicals from Cajamarca to adopt their understanding of the concept; or they tolerate a difference in understanding and in so doing maintain a kind of two-tiered double parallelism. That is, foreign Christians working in parallel with local Christians who think differently about how to live in parallel to the wider population. For example, foreign missionaries and Christians from Cajamarca may worship and work to influence wider society in Cajamarca together despite holding different ideas about what is appropriate for women, how to participate in cultural events, and the acceptability of alcohol consumption.

Even though the language of being "in the world but not of the world" might seem to imply a kind of geography or spatiality of participation, interviewees primarily understood this idea not in spatial terms but spaceless moral terms. When asked more generally about what (if any) parts of the world of Cajamarca they actively try to maintain a separation from, answers were invariably connected to moral objections. What the participants felt morally unable to participate in, however, differed significantly. Aron and Carola stated Carnival as the primary thing that they felt they needed to maintain a separation from. Cristhina, a missionary from Brazil, stated witchcraft or pagan rituals were more common in Cajamarca than in her region of origin and she and her husband refused to engage in them. Naomi, an US citizen living semi-permanently in Cajamarca listed the normalisation of cheating, lying to avoid embarrassment, demeaning jokes among friendgs, littering and fear or shame-based teaching strategies as things that she tried to separate herself from. John, a US citizen who led a Calvary Chapel Bible College for 14 years in the rural outskirts of Cajamarca, said that corruption (for example, bribery and using false or misleading documents) is something he encountered in Peru and always refused to engage in, even if it risked him receiving arbitrary fines or generating other administrative problems.

Carys explained that the use of corporal punishment (which is still commonplace in Peru though technically illegal) was not only something that she wanted to be entirely separate from, but also something that raised sometimes distressing questions about how to respond in terms of safeguarding policies and practices. Arthur took a pragmatic approach to a similar concern. He stated that while in his own community of origin he feels equipped and morally obliged to actively engage in the prevention of gender-based violence or other kinds of violence or intolerance, if he witnessed something he found personally unacceptable while in a foreign country he would avoid getting involved, deferring to locals to respond. Personally, I desire to remain separate from class-based discrimination (discussed below), administrative corruption, corporal punishment of children *and adults* (which is also common through community organised vigilante groups called *las rondas*) and gender stereotypes that encourage violence, all of which are part of my everyday experience in Cajamarca as a long-term resident. No missionaries suggested that they view Cajamarca as a more morally corrupt place than their country of origin. They did express, however, that the moral challenges and dilemmas they encountered were *different*, which required an ongoing process of reflection and adaptation to respond.

All participants have sought to connect and integrate into the world of Cajamarca and have generally made significant personal sacrifices to do so; they seek *togetherness*. The parallel desire to retain a moral separation from certain aspects of the world of Cajamarca mentioned above can be quite distressing because it can create relational barriers. Missionaries can feel required to temporarily suspend and constantly reevaluate their moral convictions in order to belong in and accept their new residence.

Naomi described this feeling in detail with reference to demeaning jokes (e.g. sarcastic or insulting nicknames or insults based on some point of insecurity), which are very common among (especially male) friendship groups in Peru. Many Peruvians that I know have described this as part of their culture, something that differentiates Peru from its neighbours (like Bolivia and Ecuador) in a way that they either feel proud (or at least unashamed) of. The ability to make these jokes indicates feeling safe for some Peruvians as it is something they would do exclusively with friends and family or neighbours from *el barrio* (their neighbourhood). Formal, unfamiliar contexts or contexts in which they feel insecure about their belonging preclude the freedom to make these jokes or use these nicknames. Having said that, while *making* the jokes at others' expense indicates the person making the joke feels safe, the person being demeaned often dislikes, feels insecure, or even resents the jokes being made. Naomi described her moral struggle to navigate this engagement:

> Demeaning jokes. The separation is interesting because sometimes I'll make comments but mostly out of discomfort I'll just resort to nervous laughter with I think just encourages it. [...] I would like to be more intentional about not just laughing, but either being

uncomfortably silent or calling it out. Which is also awkward because how much of it is a cultural thing? It's very very confusing.

Having grown up in Scotland, the kind of cutting, personal humour that is common in Peru feels less alien to me; rather it is a point of cultural connection. What level of moral compromise to make in respect for the culture of Cajamarca or pragmatically to establish and maintain relationships with people can be unsettling and feel very consequential. The very desire to maintain a separation from some aspect of the world of Cajamarca implies an ongoing, largely unresolved process. It is in this moral realm that my sense of identity and belonging are most deeply challenged as a missionary. Several others described similar experiences of intense emotional and psychological investment when addressing these moral tensions.

In conclusion, the desire for separation from "the world" and/or morally unacceptable aspects of the world of Cajamarca is not homogenous among missionaries. It is not static or indifferent. It is an ongoing, consequential, layered and complex give and take. This is not to say missionaries are necessarily moralistic. For all participants, engagement in these moral struggles comes at a personal cost and is at least ostensibly motivated by the Christian ideals of God's unconditional love and the command to love your neighbour.

Parallel Geographies of Class

Another set of complex parallelisms that emerged were related to class. Class structures that produce geographical parallelisms exist in virtually every society; however, the particular dynamics and geographies of class in Cajamarca can be quite different to those in missionaries' countries of origin. North American and European missionaries are generally assumed to be wealthy, highly educated or "cultured" (*culto/a* or *educado/a*) by people from Cajamarca regardless of their social location, wealth, or level of education in their country of origin. Navigating these new configurations of class can be confusing and disorientating. Carys, for example, described an experience where the kitchen at church ran out of cooking oil (which costs about US $1 to buy from the shop across the street) and the pastor (who would normally buy the oil) was not present. This situation generated some distress and confusion both on the part of the cooks and Carys who could not understand why it would be a problem or a source of conflict to work out how the oil could be bought so that food preparation could continue. It was not clear if any kitchen volunteers had the money immediately available or if there were relational or class dynamics at play in deciding whose responsibility it was and who would be reimbursed if someone volunteered to buy it before the pastor arrived. Carys did not get an immediate resolution to this confusion.

Places like shopping malls or airports can make North American and European missionaries feel "at home" because they are similar to places in their country of origin. The Starbucks, KFC and H&M in Cajamarca, for example, look and feel the same as they do in the USA and in England. Carys described feeling less out of place there than in the poorer areas of Cajamarca:

> When I went to the shopping centre, the mall, that was drastically different to the Cajamarca I was used to seeing; people had nice clean clothes, more like what I see in my culture. I didn't stand out as much there.

For Cajamarca residents, these places are considered quite middle-class and can feel unwelcoming or out of reach to poorer people. It is not uncommon to see signs that read "discrimination is not permitted on these premises." The discrimination implied by these signs is primarily class-based discrimination and the need for the signs is itself evidence that this kind of discrimination is a risk for poorer people. The first year I lived in Cajamarca, I vividly remember wanting to spend time in *El Quinde*—the only shopping mall at the time and since renamed *Megaplaza Cajamarca*—to feel temporarily back in the "comfort and safety" of modern Western society. I also vividly remember going with a group of adolescents and leaders from our church in the urban periphery to the Pizza Hut at *El Quinde*. The adolescents were excited and fascinated by the whole experience. This experience revealed the enormous gap between our experiences of the same place. What can feel homely for Western missionaries can feel alienatingly (or excitingly) middle-class to the people they serve if they serve in poorer communities. Figure 3 compares the poor urban periphery of Cajamarca to the historic centre and shopping mall and Fig. 4 compares the poorer agricultural outskirts of Cajamarca with the airport (which is also located on the city's outskirts).

Missionaries can leverage this social status to build prestige or appeal to their work. Aron and Carola are clear to include the American nationality of visiting expositors and leaders in the publicity for the events they organize. They also state that their work seeks a kind of pure Christianity without "culture", which would "get in the way" of advancing the Kingdom of God.

> The name of our ministry is Seeds of Love Missions so our goal our mission is to sow seeds of love wherever we go. That people would know the love of God versus the love of doctrine, the love of religion and the love of culture. Sometimes these are very juxtaposed.

An example of "culture" as an obstacle would be the expectations for women to wear a skirt mentioned already above.

Leveraging North American prestige and simultaneously assuming that is possible to exist beyond culture fits uncomfortably well into Escobar's description of the One World World, viz.: the colonial universalizing of American culture in the guise of pure true religion. On the other hand, however, seeking a pure, cultureless Christianity is also something that Aron, Carola and their congregation the USA do in contexts that have a much less obviously postcolonial subtext. This universalising and idealistic posture might be better understood as having its deepest roots not in colonialism or American exceptionalism, but in Platonic idealism, the influence of which in Western Christianity has been vigorously debated and critiqued in recent decades (Ruether, 1992). More recently, Jennings has cogently argued that this kind of theological posture may have its deepest roots in the initial struggle in the early church to navigate the relationship between Jewish and Gentile Christians (Jennings,

Fig. 3 *Top left*: The church where Carys volunteered. *Top right*: The urban periphery of Cajamarca. *Bottom left*: The historic centre. *Bottom right*: Megaplaza Cajamarca [formerly *El Quinde*] (Sources: Olga Cueva and author)

2011). A "One World World" attitude may not, therefore, be exclusively the result of European colonialism, but also ancient and ongoing philosophical and theological struggles surrounding the role of race and culture within Christianity since its foundation.

Moreover, locals from Cajamarca are under no illusion that Americans exist beyond or outside culture. To assume that a universalist or transcendent/idealistic view of culture on behalf of missionaries would override people from Cajamarca's own sense of the world and of difference would drastically underestimate their creative capacity for cultural subversion, which has a long history. As mentioned above, Cajamarca is known historically as the place of encounter between European and Andean cultures as a centre for Carnival celebrations nationally. A central figure of Carnival in Cajamarca is the *Ño Carnavalón*, an effigy pictured in Fig. 5 that is wheeled around the city for three days and then buried in a satirical funeral with men dressed as women acting as professional mourners (Fig. 6).

The figure of the *Ño Carnavalón* is connected to *Momus*, the Greek god of satire and mockery. Cecilia Lopez, an academic at the University of Piura, in a region neighbouring Cajamarca, has described this connection in the following way:

> The famous Ño Carnavalon or King Momus, whose name comes from the Greek Momus, a god who represents sarcasm, mockery and irony, is the king of the carnival party. This is, then, the reason for which the King Momus becomes the centre of attention throughout the festival because it satirizes and ridicules the behaviours of the people in the area. Once buried, the reading of his will reveals harsh criticisms generally directed against the

Fig. 4 *Top left and top right*: Traditional smallholder family farms in the agricultural outskirts of Cajamarca. *Bottom left*: Church community meal in the agricultural outskirts of Cajamarca. *Bottom right*: Cajamarca airport (Source: author)

> principal events and people of the region where the festival is celebrated. (Lopez, 2015; author's translation)

Joyfully, mischievously, and sometimes maliciously satirising and subverting the expectations of the powerful in the face of outside oppression is one of the symbolic heartbeats of Carnival in Cajamarca.

Unsurprisingly then, people from Cajamarca can and do also use the prestige of Western missionaries strategically for their own ends, sometimes unbeknownst to the missionaries themselves. On several occasions I have been invited by people from Cajamarca to participate in events where one of the justifications given was that my presence would lend the event *realce* ("importance", "lustre", "brilliance" or "splendour"). One of my friends who had made the invitation once explained, "I don't know if you have noticed but when foreigners are involved it makes us look more important" [authors paraphrased translation].

These experiences have made me reflect critically on my engagement with class as I was growing up in Scotland. As a teenager it had felt unproblematic to use class-based slurs and to this day I experience the effect of class-based accent bias in the UK kind of viscerally. I can feel patterns of prejudice activated in my thoughts and emotions there. In contrast, when in Cajamarca, the arbitrariness and injustice of class-based marginalization produce indignation in me. I feel much more

Fig. 5 The entrance of the *Ño Carnavalón* in the streets of Cajamarca (Source: author)

comfortable in places associated with extreme poverty in Cajamarca than I do in Scotland. In other words, settling in a different culture has made new affective responses to class available to me. Carys described a similar experience of class dynamics in Cajamarca:

> Hearing the way that [middle class] people [from Cajamarca] would speak about the people I was serving [in the urban periphery] and hearing the judgement and disregard and how they would look down on them and see them as bad people reminded me that the people I was serving weren't representative of all of Cajamarca.

All participants in this study move with remarkable ease and apparent comfort between different types of social space in Cajamarca, moreso than what is common for people from Cajamarca. Cristhina and her husband visit the local prison weekly. Naomi and Carys described feeling deeply at home playing, eating and learning with children in a church in the urban periphery of Cajamarca, a neighbourhood marked by poverty, violence, and substance abuse. Aron and Carola organise events in one of the most expensive hotels in the city's centre and also spend significant amounts of time in some of the poorest rural communities. It would be unusual, if not somewhat shocking or suspicious, for people from Cajamarca with a similar level of social prestige as missionaries to frequent the prison, the urban periphery, or poor rural communities. By building meaningful and trusting relationships across

Fig. 6 Men dressed as women satirically "mourning" the burial of the Ño Carnavalón (Source: https://www.infobae.com/peru/2025/02/26/carnaval-de-cajamarca-cual-es-el-origen-del-no-carnavalon-la-figura-principal-de-la-celebracion/)

these local geographies of class, missionaries participate in the subversion of otherwise quite stable structures of class discrimination.

It is not clear if and how these experiences affect the way that these missionaries engage in class when back in their country of origin. However, being a missionary can create new opportunities for self-expression and for re-formulation of personal identity. It can afford people the opportunity to reassess and reposition themselves with respect to class dynamics.

Attempts by missionaries to cross class boundaries are not always successful. John, an American missionary whose organisation (Calvary Chapel) built a bible college in Llacanora (a rural community on the outskirts of Cajamarca) chose this location because the staff wanted international students in their school to be able to connect with the land and the local community through farming. Although they had not initially planned to build a wall around their property, it became urgent because of constant robberies. Meaningful and interdependent relationships were still fostered with the surrounding community; however, this task took many years and the wall that separated the college's property from the surrounding farms gave material expression to the division and parallel existence of the different groups.

In conclusion, missionaries encounter new class landscapes when they travel to another country. They can reify and benefit from unjust class dynamics; their class privilege can be exploited by people from Cajamarca for their own ends; and they can have new social and affective freedom to transverse or even subvert class-based and exclusionary parallel geographies with varying levels of success.

The Shared Table

One locus emerged throughout this study as having particular symbolic and practical importance in the negotiation of parallel worlds in Cajamarca: food, food culture, *the table*. These were mentioned unprompted by all interviewees and were often an important piece of what the missionaries wanted to talk about when discussing (dis)connection with the world of Cajamarca.

Sharing the table can be understood as a place of deep, intimate encounter with the Other with the (sometimes illusive) goal of togetherness. It was understood by the missionaries in three aspects. Firstly, sharing the table is *a universal human practice*. Naomi commented that eating is the most public basic human need which in itself makes it a point of human connection. Arthur described a similar idea like this:

> A place where I feel included and that is very symbolic is the table. When I am at the table with people form Cajamarca, I feel part of Cajamarca. We partake of the same food, we partake in the same human need to be nourished. We partake of our humanity. It is a very impactful space where I can see myself as being from Cajamarca.

Secondly, interviewees expressed that sharing the table is *a Latin-American practice*. Arthur said that "we are blessed as Latin Americans because in some way, perhaps through our indigenous ancestry, the table is a place for everyone" [author's translation]. Terry, the pastor of Aron and Carola's sending church in the US who visits Peru regularly, also expressed being impressed by the general food culture in Peru, highlighting the excellence and pride with which it is prepared and served:

> Terry: It's the way [food service] is set up is just different to what I've ever experienced[...]You know, I feel like every place I've been its like they cook with their heart not just out of a restaurant. They love to be a blessing, they want you to enjoy that food.
> Aron: It's like they express themselves with their food

Thirdly, sharing the table is understood as *a Christian practice* by many participants. Arthur explained the symbolic and relational value of the table by referencing stories from the Gospels. Luke 15, for example, contains three stories of someone or something lost being found (a coin, a sheep, and a son). Arthur noted that in all three stories the recovery of what was lost is celebrated at the table. The narrative where Saint Peter is reconciled with Jesus after having denied him three times takes place over breakfast. The table, for Arthur, is a place of relationship building at the spiritual level. Naomi noted how many meals are specifically mentioned in the Bible, particularly in the life of Jesus. She also highlighted the fact that the Eucharist (also known as Communion or The Lord's Supper) can, and perhaps should, be understood as a shared meal, not just a religious rite.

Terry observed a connection between food culture in Peru and their spiritual expression:

> The way they present food with great pride and want you to enjoy it, it's the same thing spiritually. They are unafraid and unashamed to worship their God, as Christians to let their heart be known to the Lord. [...] Just in the way that I am hungry for the delicacies, they are hungry for the Word of God.

Part of what makes sharing the table a spiritual practice for Naomi is that it requires a level of ecological humility: "The humility of acknowledging that food is a gift from God rather than thinking that we figured it out. Receiving from God is one part of it."

John's experiences revealed a clear example of this. While they had initially planned to engage in agriculture at their bible college to facilitate some self-sufficiency and to keep costs down, it ended up creating bridge-building *dependence* on neighbouring farmers' local knowledge. John describes large celebrations where local food was prepared as "real bridge builders" and moments of deep cultural exchange. Arthur expressed a similar idea suggesting that the table as a Christian practice requires a commitment to a kind of radical equality and inclusion based on the universality of Jesus' call:

> Where there is a culture of The Table, differences are not thought about. It is a table where there is no head. At our table, we do not look at who has a place and who doesn't. The Table belongs to Jesus. And if Jesus calls everyone, I cannot complain about who sits next to me. If Jesus has died for all, all humanity, who am I to say who does or does not have a place at the table.

Naomi expressed the spirituality of the table through the negative sense of how a food culture can express selfishness:

> there is a way to eat and have a meal that is very selfish and there is a way to eat and enjoy food for yourself but also very selflessly. What you eat, how much you eat, how you interact with the other people eating, how long you eat for. You imagine someone stuffing their face alone versus a family really enjoying each other's presence and having worked together to get the food on the table, to be together, to clean up, whatever it is, it's a whole social, loving encounter that you also get to enjoy and receive from.

As a life-long vegetarian, Carys' experience with food culture in Cajamarca was much more fraught. Vegetarianism is uncommon and sometimes misunderstood. It sometimes led to uncomfortable situations where Cary's refusal to eat a particular meat-based food risked relational damage with hosts providing the food. However, this dynamic reinforces the idea that the table is place of deep encounter with the Other. By sharing the table, both Carys and her hosts were forced to wrestle through a process of learning to navigate, accept and respect (though not necessarily share) value systems that felt alien. Through a process of spiritual discernment, Carys ultimately arrived at the conviction that as a disciple of Jesus it would be more appropriate sometimes to accept and eat meat provided from a host as an act of loving sacrifice than to refuse it. Sharing the table is an opportunity for both host and guest to encounter the Other and to choose to compromise their own desires and sometimes even their deeply held moral convictions to connect with the Other.

In summary, the shared table can be a place of deep intimate encounter with the Other, their worlds and their shared embodied existence with the land and with the divine. To describe it as an encounter recognises the potential for discomfort and tension but also deep connection, belonging, acceptance and personal transformation. The table is where our differences and our shared existence cannot be hidden but must be acknowledged and addressed *together*.

Conclusion

In conclusion, I offer a reflection for three different audiences. To social and political theorists, I suggest that although this study recognises the problem of imperialist and capitalist reductionism that Escobar's One World World narrative highlights, encounter and togetherness are more fruitful avenues than radical onto-epistemological pluralism in the search for justice and ecological wholeness. The radical pluralism of multiple worlds with no underlying unifying reality would, in fact, *exclude* the potential for encounter and togetherness. Given the profoundly unjust nature of the encounter and coexistence of Euro-American culture and indigenous cultures of the Americas over the last five centuries, radical pluralism is an alluring idea. However, it is difficult to imagine how life in Cajamarca would be better or more just for either locals or missionaries if such a radically pluralist posture was adopted such that encounter and togetherness would become unattainable. Missionaries actively wrestle with the one world/many worlds tension and many kinds of parallelism in an ongoing way in their daily lives in deeply complex, layered, personally significant ways. The interaction of morality, theology and class are important components of this process and the process of generating change. The potential for change is what makes more just futures possible.

To geographers, I suggest a closer consideration of the shared table as a locus for the (re-)creation, maintenance, dissolution, and transformation of social difference in the context of migration. For those who study migration, the shared table (or lack thereof) invites a consideration of the quality and nature of the *encounter* of different subjectivities in the everyday embodied experiences of people in different social locations. It is both a world-making practice and a world-sharing or world-negotiating practice that can impact both the migrant and her/his destination community. For those who work towards de-centring the human, the shared table may also be a fruitful avenue for building an affective awareness of our embeddedness in wider ecologies and socio-environmental assemblages.

To theologians I suggest that the shared table reaffirms the contextual and lived nature of theology. Contextual theology and lived theology as theoretical streams in the discipline have brought the contingent and embodied nature of theology to the attention of mainstream theological scholars. To this it may be important to add that contexts and lived experiences *change* through the *encounter* with difference (through migration or mission, for example). There exists the need to consider this encounter with difference in the construction and maintenance of contextual and lived theologies. The discussion of the shared table points to the potential for lived theologies of food to contribute to a re-discovery of humanity's ecological embeddedness.

Finally, there is a need for academic contributions in these areas from religious scholars from within different religions. Within academic geography, there have been several decades of discussion surrounding the need to recognise the positionality of the researcher and their role as a subject in the construction of geographical knowledge (Rose, 1997). Despite this, theological frameworks and religious commitments have generally been treated as an object of study within academic geography, not a lens through which researchers themselves view and understand their world and the worlds of others.

Acknowledgements I would like to thank who were willing to be interviewed for this project: Carola, Aron, Terry, John, Carys, Naomi and Cristhina. I would also like to thank the people of Cajamarca who received us into their homeland.

References

Andolina, R., Laurie, N., & Radcliffe, S. A. (2009). *Indigenous development in the Andes: Culture, power, and transnationalism*. Duke University Press.

Bennett, J. (2009). *Vibrant matter: A political ecology of things*. Duke University Press.

Berry, W. (2002). The whole horse. In E. T. Foygle (Ed.), *The new agrarianism: Land, culture and the community of life* (pp. 63–79). Shearwater Books.

Berry, W. (2010). *What are people for?: Essays*. Catapult.

Brueggemann, W. (1991). *Interpretation and obedience: From faithful reading to faithful living*. Fortress.

Brueggemann, W. (2000). The liturgy of abundance and the myth of certainty. In P. D. Miller (Ed.), *Deep memory, exuberant hope* (pp. 69–75). Fortress.

Davis, E. (2009). *Scripture, culture, agriculture: An agrarian reading of the Bible*. Cambridge University Press.

de la Bellacasa, M. P. (2017). *Matters of care: Speculative ethics in more than human worlds*. University of Minnesota Press.

de la Cadena, M. (1998). Silent racism and intellectual superiority in Peru. *Bulletin of Latin American Research, 17*(2), 143–164. https://doi.org/10.1016/S0261-3050(97)00085-5

de la Cadena, M. (2015). Uncommoning nature. *E-Flux Journal 56th Biennale, 65*, 1–8.

Escobar, A. (2018). *Designs for the pluriverse: Radical interdependence, autonomy, and the making of worlds*. Duke University Press.

Escobar, A. (2020). *Pluriversal politics: The real and the possible*. Duke University Press.

Fergusson, D. (2011). Creation. In I. A. McFarland, D. A. S. Fergusson, K. Kilby, & I. R. Torrance (Eds.), *The Cambridge dictionary of Christian theology* (pp. 120–122). Cambridge University Press.

Greene, S. (2006). Getting over the Andes: The geo-eco-politics of indigenous movements in Peru's twenty-first century Inca empire. *Journal of Latin American Studies, 38*(2), 327–354. https://www.jstor.org/stable/3875502

Jennings, W. J. (2011). *The Christian imagination: Theology and the origins of race*. Yale University Press.

Jennings, W. J. (2019). Reframing the world: Toward an actual Christian doctrine of creation. *International Journal of Systematic Theology, 21*(4), 388–407. https://doi.org/10.1111/ijst.12385
Keller, C. (2021). *Facing apocalypse: Climate, democracy, and other last chances*. Orbis Books.
Lopez, C. (2015). El Ño Carnavalón y la yunza. In *Castellano Actual de la Universidad de Piura* [online]. Retrieved September 4, 2025, from https://www.udep.edu.pe/castellanoactual/el-no-carnavalon-y-la-yunza/
Medina, N. (2021). (De)ciphering mestizaje: Encrypting lived faith. In M. H. Diaz (Ed.), *The word became culture MLA 8th edition (Modern Language Assoc.)* (pp. 47–55). Orbis Books.
Rose, G. (1997). Situating knowledges: Positionality, reflexivities and other tactics. *Progress in Human Geography, 21*(3), 305–320. https://doi.org/10.1191/030913297673302122
Ruether, R. R. (1992). *Gaia and god*. Harper San Francisco.
Singh, N. M. (2018). Introduction: Affective ecologies and conservation. *Conservation and Society, 16*(1), 1–7. https://www.jstor.org/stable/26380571
Sugirtharajah, R. S. (2022). Introduction: The Bible, empires, and postcolonial criticism. In R. S. Sugirtharajah (Ed.), *The Oxford handbook of postcolonial biblical criticism* [online]. Retrieved August 26, 2025, from https://doi.org/10.1093/oxfordhb/9780190888459.013.37
Tsing, A. L. (2015). *The mushroom at the end of the world: On the possibility of life in capitalist ruins*. Princeton University Press.
Wallace, M. I. (2020). *When god was a bird: Christianity, animism, and the re-enchantment of the world*. New York Fordham University Press.
Wirzba, N. (2007). *The paradise of god: Renewing religion in an ecological age*. Oxford University Press.

Uba Talaiashke: Creolization and Pre-Christian Language in Choctaw Hymns

John DePriest

Introduction

Much of the legacy of Christianity for indigenous communities in the Americas is defined by a history of harm. Still, many Indigenous people and communities have embraced Christianity and have even adopted it as a central feature of tribal identity. The Choctaw Nation of Oklahoma exemplifies this complex relationship. Many Choctaw students sent to religious schools were victims of abuse and neglect, and missionaries actively sought to end traditional cultural practices they saw as heathen—and in some cases succeeded. Conversely, missionaries also helped develop a written Choctaw language and brought an American-style education system into the Nation before and after removal, leading to expanded economic opportunity. Many Choctaws, especially community leaders, converted to Christianity beginning in the 1820s and 30s, and today the Choctaw Nation is officially a Christian nation, even though many Choctaws do not identify as Christian and never have. For those who have adopted Christianity, it is often a central part of their identity as Choctaw people. Today, church-based hymn singings are opportunities for community building and affirmation of Choctaw identity (see Fig. 1 for a map of Choctaw hymn-singing sites). The churches and community centers where tribal members gather to sing represent a parallel universe within the world of American Protestant Christianity where Choctaw identity is proudly expressed through a unique creolized tradition of Choctaw Christianity and hymnody that connects tribal members to their ancestors and community.

J. DePriest (✉)
Center for Global Education, Tulane University, New Orleans, LA, USA
e-mail: jdepries@tulane.edu

S. D. Brunn (ed.), *Transdisciplinary Research Across Knowledge Worlds*,
https://doi.org/10.1007/978-3-032-19491-6_8

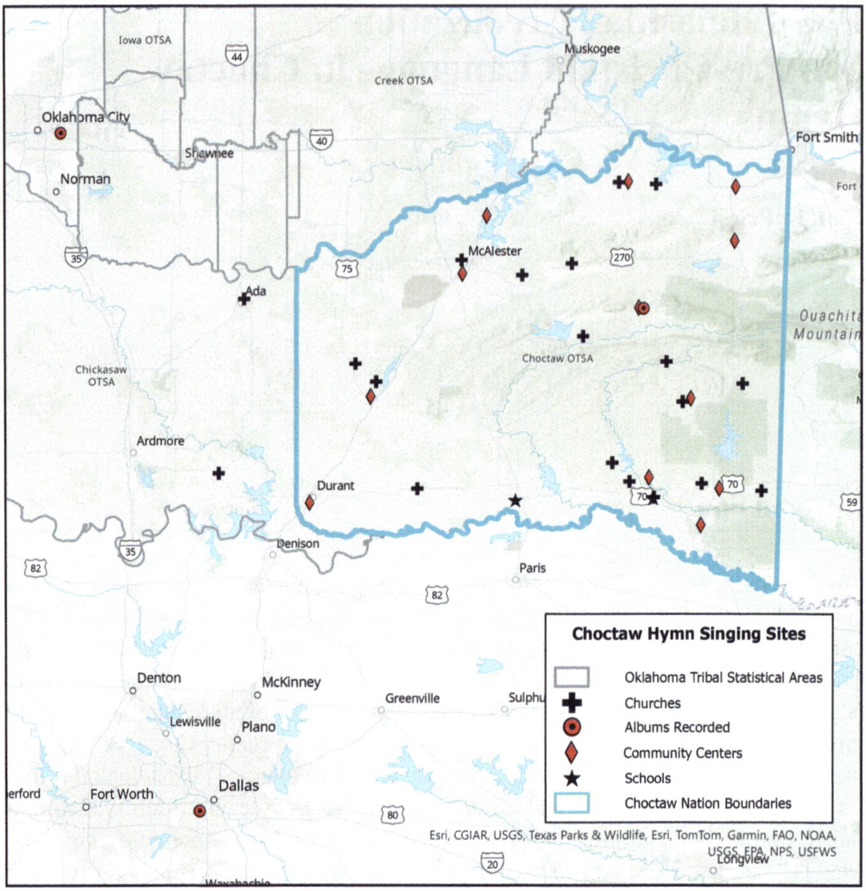

Fig. 1 Some community gathering places where hymn-singings occur in and around the Choctaw Nation, including Choctaw Nation community centers, churches (including three where albums of hymns were recorded), and mission schools (map created by the author from public announcements of hymn-singings and recorded albums)

Creolization

Creole cultures in the American South, the Caribbean, and elsewhere provide a model not only for understanding Choctaw Christianity, but also the development of the specific cultural tradition of hymn-singing in the Choctaw language. Spitzer (2003: 58) describes the process of cultural creolization as the "development of new traditions, aesthetics, and group identities out of combinations of formerly separate peoples and cultures—usually where at least one has been deterritorialized." He goes on to say that "[c]reolized societies conjoin two or more formerly discrete cultures in a new setting to create a social order in which heterogeneous styles, structures, and contents are differentially preserved while becoming wholly constituted in and adapted to new circumstances with new and multifaceted meanings" (Spitzer, 2003: 58–59). In this culturally creolized context, traditional forms can still hold meaning and usually do. In fact, these traditional forms are often revered and emphasized, especially in the creation of new art. In the case of Choctaw Christianity, the "heterogeneous styles, structures, and contents" include traditional Choctaw language, music, community, shared history, and religious beliefs merged with influences from Anglo and African American Protestantism. Choctaw hymns are an art form that emerged out of this creolization process, combining aspects of traditional Choctaw community gatherings, theology, and music making with the sacred harp singing tradition popular in the nineteenth century. The formation of this creolized genre, identity, and culture occurred in the context of the deterritorialization of the Indian Removal Act beginning in 1830, when the Choctaw Nation became the first nation forcibly relocated to what later became known by the Choctaw name Oklahoma.

While many refer to the ritual and cultural outcomes of religious creolization with the blanket term of syncretism, today many scholars and religious practitioners reject the term. For some scholars, distinguishing between syncretic and non-syncretic religions becomes an exercise in futility because "no culture is an island, and … borrowing and mixing have always gone on" (Lindenfeld & Richardson, 2012: 4). For practitioners, the description of syncretism may suggest that their religion is less "pure" or somehow corrupted. In Brazilian Candomblé, some leaders argue that their religion remains consistent with its traditional African origins and have called to purge their practices of European elements (Johnson, 2002). For many Christians in the United States whose religious practice has been influenced by African or Indigenous traditions, the suggestion that their religion is less Christian would be quickly rejected. On the other hand, some scholars find that the term is useful for describing religious intermixing, since many people have practiced syncretic religions "even when they were unaware of it or sought to deny it—as they not infrequently did" (Lindenfeld & Richardson, 2012: 4). Considering that religious belief is only one aspect of cultural practice along with music, identity, and language use, a broader cultural framework is required and used here. While a case could be made that Choctaw Christianity is syncretic because it includes pre-Christian elements that have merged with its belief system and practice, I prefer the term creolization for both its explanatory power and its applicability beyond religion.

In their history of interaction with missionaries and conversion to Christianity, the Choctaw experience shares commonalities with many other Indigenous and African-descended groups in the Americas. Neylan (2003: 17), in her work with Tsimshian Protestantism, describes conversion as "rarely absolute" and states that the construction of religious beliefs depends on cultural context and necessitates "a blurring of boundaries between spiritual systems." Similarly, in Brazil, some church leaders encouraged the syncretic association of traditional and Christian faiths, viewing it as a "positive step on the path towards conversion" (Johnson, 2002: 71). Samson, in discussing contemporary Evangelical conversion among the Maya in Guatemala, highlights the need for individuals to see themselves "as still connected with their ancestors" who may have had very different theologies. For the Kiowa in Oklahoma, Lassiter (2001: 242) notes that "while Indian churches share basic doctrine with their denominational counterparts in other American communities, their obvious differences in membership, individual church histories, and the use of tribal language in speech and song have brought about a unique Christian practice." In these cases, as well as for Choctaws, Christianity has become a key component in negotiating and maintaining community identity in a changing world. Despite sharing cultural commonalities and proximity with their neighbors, these communities have distinct experiences, theologies, traditions, and motivations for engaging with Christianity. While many aspects of worship are shared with their neighbors, these religious traditions remain separate, operating in parallel to those of Anglo-American Protestantism both within and outside the bounds of the reservation.

Traditional Religious Context

Aside from the use of the Choctaw language and the Choctaw tribal membership of churches, isolating the Choctaw elements of contemporary Choctaw Christianity can be challenging for a few reasons. First, little was written about Choctaw beliefs prior to the nineteenth century, at which point Native Southeasterners had experienced varying levels of contact with European colonizers for over 250 years. Second, Choctaw people were documented as having a "great aversion" to sharing religious beliefs or traditions, especially with outsiders (Wright, 1828: 178). The potential for ridicule and dismissal from outsiders in response to sharing traditional knowledge likely played a role in this aversion. Third, many traditional Choctaw beliefs were similar to those of the Abrahamic religions, so it can be difficult to distinguish between older elements of traditional religion and the religion of the colonizers. In these cases, though, there are often linguistic elements that persisted through the conversion to Christianity that were used before and after, both by missionaries and Choctaws themselves that suggest the preservation of traditional religious beliefs.

When colonizers arrived in the Southeast, they found varied and diverse societies with deep, long-standing religious traditions. Many were the descendants of Mississippians, whose cultural symbols they inherited and continued to use. Europeans found similarities between the symbolism and belief systems of Native Southeasterners and Abrahamic religions. Some of the most expansive documentation from the eighteenth century came from James Adair (1986 [1775]), who considered Southeastern tribal communities, particularly the Cherokee, Muskogee Creek, Choctaw, and Chickasaw, to be the "lost tribes of Israel" and devoted 230 pages of his *History of the American Indians* to 23 arguments in favor of this erroneous argument. These cultural similarities, as well as a comparative enthusiasm to adopt European goods and market economies, led newcomers to call these communities "civilized tribes." Also due to these pre-existing commonalities, conversion was likely more of a modification of traditional beliefs rather than the wholesale rejection of the traditional in favor of the modern.

One of the most obvious commonalities between traditional Southeastern religion and Christianity is the prominence of cross symbols, especially the cross-in-circle configuration in Mississippian symbolism. While it is sometimes thought to represent the Earth through reference to the cardinal directions, or origins of winds, (Gatschet, 1884), it is most often associated with the sun (Fundaburk & Foreman, 2001), a symbol of the creator figure the Choctaw people historically called *Hvshtahli*, among other names. Some outsiders, such as missionary Alfred Wright, interpreted references to *Hvshtahli* as representing sun worship, but the sun was not the deity, rather the medium through which *Hvshtahli* oversaw creation (1828). Byington et al. (1915) defines this name as "the Great Spirit; the governor of the world, whose eye is the sun." *Hvshtahli* had a strong association with fire, which was considered the terrestrial partner or offspring of the sun. Fires were often built around two central crossing logs that formed the cross-in-circle shape. Indeed, Lankford (2004) considers the cross-in-circle shape a two-dimensional depiction of a three-dimensional phenomenon: that of the sun positioned directly above the crossed logs of the fire, since fire was considered the "sun's mate" (Wright, 1828) analogous to Christian connections between God the Father and the Holy Spirit, which despite having separate names are elements of a single deity. Most Southeastern cultures were similarly monotheistic prior to contact with Europeans. According to Cushman (1899: 362), Israel Folsom, an early and prominent Choctaw Christian, mentions that previous generations believed in "a Great Spirit, and that he possessed super-natural power, and was omniscient, but they did not deem that He expected or required any form of worship of them. . . All they felt was a dread of His attributes and character, made manifest to them by the phenomena of the heavens." Despite the lack of regular worship, this existential dread likely found a complement in the salvatory anxiety underpinning the Calvinist theology of early nineteenth century Presbyterian missionaries.

One of the other traditional names for *Hvshtahli* is *Ʊba P̱iki* (Wright, 1828: 179), which means literally "our father above." Adair recognized a phonetic correspondence between the Aramaic word for "father," *abba*, and the Choctaw *vba*, meaning "above," and to his mind, this linguistic coincidence suggested shared descent (Adair, 1986 [1775]). Swanton (1931: 197) found this term to be cognate with similar words across several Southeastern language communities: "The name of God among the Louisiana Choctaw, Aba, 'above,' is evidently identical with, or abbreviated from, a purely aboriginal designation, almost identical with one term used by the Chickasaw, Creeks, Chitimacha, and Atakapa." He does not say which terms he is referring to in the other languages, but these languages come from three different families so having a cognate suggests this term may have come from an earlier shared religious system. Wright (1828) claims that Choctaws learned *Ʊba P̱iki* "from the whites" without providing any rationale as to why. He may be referring to the *P̱iki*, or "our father" element, since there is no evidence of Indigenous Southeasterners referring to their deity as "father" prior to conversion. Wright may have been wrong about the origin of this term, but if not, *Ʊba P̱iki* could be an early indication of a creolizing religious system, with Choctaws taking elements of Christianity, i.e. the conception of a heavenly father, and combining it with a name for their own deity.

Conceptions of the afterlife are an additional area where traditional Choctaw beliefs and Christianity coincide. In pre-contact Mississippian religion, the afterlife was conceived as existing in a land above, and to reach it one would have to travel along the path of souls, the spiritual manifestation of the celestial Milky Way (Lankford, 2004). While much of the cosmological detail surrounding this journey was lost between the sixteenth and nineteenth centuries, some knowledge of the overall journey, including the destination, remained. Wright (1828: 182) extensively describes the *shilup i̲ yakni*, or "land of the spirits" that many traditional Choctaws believed was their final destination. The description bears great similarity to that of some portrayals of Christian heaven as an eternal life in paradise. Here, the deceased would experience perpetual youth, good food, and warm weather, while death was conceived of as the "shedding of the external covering of the *shilup* [spirit], as the snake sheds his skin" (Wright, 1828: 182). Nothing was apparently required of a person during their life in exchange for this afterlife; it was guaranteed to everyone unless they had committed a particularly grievous murder. For these murderers, a second place existed, within sight of the first, where they were denied all of the pleasures of the other place. On their passage to the *shilup i̲ yakni*, "they were unable to find *the bright path*, which led to the land of the blessed" (Wright, 1828: 182, emphasis added). Unlike Christianity, this outcome was not the result of a decision by a deity, but rather a matter of nature; however, these dual destinations with a connection to behavior during life easily map onto Christian concepts of heaven and hell.

It is important to note the use of the phrase "the bright path" in the quote above. The phrase references what many in the Choctaw Nation today believe is a "sacred way of life that God has given the Choctaw people" (Choctaw Nation of Oklahoma, 2025). The origins of this phrase likely stem both from traversing the "path of souls" along the Milky Way on the way to the afterlife as well as from the relationship with the deity whose eye was the sun during life. Wright (1828: 179) says that this term would often be referenced in a successful war or hunting speech, given to acknowledge the sun's role in helping the warriors "to find the bright path, which led them to victory, and returned them in safety to their homes." Later, early Christian Choctaw Captain Robert Folsom (1830: 156) mentions in a letter to the *Missionary Herald* that he and his brother, David, want to "throw away the wicked" and "walk in the straight, bright path." The use of this term was apparently supported by the missionaries, since Folsom's letter was edited before publication, but not apparently censored. Today, the Choctaw Nation's cultural artifacts portal is called *Hina Hanta*, Choctaw for "bright path," and in the site's description they say "[o]ur story includes all events along that path, from thousands of years ago to the current generation" (Choctaw Nation of Oklahoma, 2025). In the case of *hina hanta*, the transition from traditional theology to contemporary Christianity in the Choctaw Nation is seamless. The continued reference to and reverence of traditional concepts muddies the line between conversion and maintenance of traditional beliefs. When missionaries began translating the Bible and other religious literature into the Choctaw language, they maintained much of the traditional religious language, including the concepts of *kashofa* "cleanliness or forgiveness," *hina hanta*, *vba*, and others explored later in this chapter. The persistence of these terms suggests that for many, conversion may have been more a matter of subtly aligning concepts or adding an additional belief in Jesus rather than a wholesale transformation of religious ideology.

In this context it is important to remember the intra-community variability of religion. Choctaw historian Clara Sue Kidwell (1999: 150–151) mentions that while individual Choctaws adapted the "Christian religion on their own terms," others "found ways of reconciling the Christian God with traditional beliefs and practices, and some never became Christian." These parallel experiences within the Choctaw Nation mirror those between the Choctaw Nation and the United States writ large.

Origins of Choctaw Christianity

The origins of Choctaw Christianity go back to 1818 when David Folsom invited the first missionaries from the American Board of Commissioners of Foreign Missions (hereafter ABCFM) to come to the Choctaw Nation in Mississippi to establish a mission and school (Carson, 2003). The ABCFM was a predominantly Presbyterian group with strong Calvinist leanings. Choctaw leaders at the time, including Apuckshanubbee, Pushmataha, and Moshulatubbee as well as the next generation of leaders such as David Folsom, Greenwood LeFlore, and Peter Pitchlynn, saw education as the primary benefit of these missionaries' presence. They foresaw that education would give future generations an advantage in dealing with the white settlers arriving in their territory as well as with representatives of the US government. The first school was established at a settlement they named Eliott on the Yalobusha River (Carson, 2003). Two years later, Cyrus Byington arrived from Massachusetts to establish the missionary school at Mayhew (Debo, 1934). Byington, along with missionary Alfred Wright, would eventually develop and publish a number of textual resources in the Choctaw language, including the hymnal, a definer, and eventually, a full dictionary.

Despite the increase in missionary activity and the establishment of schools in the Nation, there was not much enthusiasm for conversion among Choctaws for the first several years. Their arrival was not the first time the Choctaw Nation had encountered Christianity; their ancestors had been interacting with European Christians for over 275 years. But despite frequent interaction with Catholic missionaries, there were relatively few converts (Wright, 1828). This lack of missionary success was likely due to an imbalance in power dynamics in favor of the Choctaws and a focus on trade and alliances over conversion (Ellis, 2022). When the missionaries from ABCFM arrived, they found that most citizens of the nation were unfamiliar with basic tenets of Christianity, aside from an awareness that the priests "had the power to send a soul ... to a good place somewhere in the regions above" (Wright, 1828: 179). For the first several years of their missions, they had little more success than the Catholic missionaries. However, when Methodist minister Wiley Ledbetter came to Mayhew in 1824, he inspired "unusual excitement" from the Choctaw attendees, particularly women, who were central powerholders and culture bearers (Kidwell, 1999). The Presbyterians dismissed this success, but Kidwell (p. 148) finds that Methodist theology was closer to traditional Choctaw values of "freedom of the will, universal access to grace and salvation, and the importance of a mystical experience of closeness to God." Starting in 1828, Methodist missionary Alexander Talley used Greenwood LeFlore's residence as a base for riding a circuit of camp meetings around Choctaw country, meeting people near their homes and using existing community gatherings as opportunities to preach. The number of converts increased, but there were other factors involved, including growing literacy, the conversion of young leaders, and changing power dynamics with encroaching settlers.

With the establishment of schools, Choctaw children began learning to read in both Choctaw and English. There was initially some debate among the missionaries about the language of religious instruction, but the Choctaw leadership and populace were adamant about learning to read in their own language, even though at the time, there was nothing to read in the Choctaw language (Carson, 2003). Byington and Wright addressed this by developing a curriculum and printed materials, including a spelling book. In this spelling book, as soon as the basics are established, the first full paragraphs consist of religious works, including "A Short Statement of the Moral Law," descriptions of heaven and hell, including who is destined for both, and "An Address to Drunkards" (Wright & Byington, 1827). In other words, as soon as young Choctaw minds learned to read in their own language, the only materials available to read were Calvinist religious tracts.

The rise of Christian leaders within the nation, especially David Folsom and Greenwood LeFlore encouraged more of the population to convert. Folsom grew up working in his father's tavern on the Natchez Trace, so he gained an early familiarity with the ways of Americans. In addition to inviting the missionaries to the Choctaw Nation, he was an early convert to Christianity and was elected Chief of the eastern district in 1825. He then aggressively campaigned to elect new reform-minded leaders in the other districts leading to LeFlore taking over in the western district. Their platform consisted largely of rejecting removal treaties while embracing the "civilizing" aspects of education and conversion to Christianity (Carson, 2003). Folsom's leadership and enthusiastic support for Christianity encouraged other tribal members to convert, leading to a period some refer to as "The Great Awakening" in the Choctaw Nation (GBS, 2017).

Folsom's ascent to power and official sanction of Christianity did not occur in isolation. Since Thomas Jefferson, the prevailing Indian policy in the U.S. had been to encourage the "civilization" of indigenous nations through education, conversion to Christianity, and the development of market economies. The expectation was that tribal governmental power would gradually erode, allowing white Americans access to tribal land holdings. Folsom's leadership sought to prevent such an outcome. He and others mistakenly believed that leading the Nation to adopt American approaches to religion, education, and economics would prevent hostilities, legal encroachment, and land-loss from the increasing number of white settlers, the State of Mississippi, and the federal government, respectively. Despite united leadership, these efforts were ultimately unsuccessful. Following the passage of the Indian Removal Act, the US Government forced the Choctaw leadership to sign the Treaty of Dancing Rabbit Creek ceding the remainder of their land east of the Mississippi in exchange for land in present-day Oklahoma, beginning the genocide the Choctaws named "the Trail of Tears," often referred to today as removal.

At the time of removal, the number of converts was still low as a proportion of the overall Nation. Missionaries would report their success to the *Missionary Herald*, and in these letters, they count converts in the single or low double digits. In 1830, one missionary reported receiving 22 converts at Hebron, bringing the total to 53 (Cushman, 1830). By this time, the ABCFM counted around 350 converts among the Choctaws, while the Methodists claimed around 1300 (Hiemstra, 1948; Kidwell, 1999). Out of the roughly 20,000 members prior to removal, this makes for a proportion of around 8%. By some estimates, over 18,000 Choctaws began the long trek to Oklahoma between 1830 and the early 1840s. Many of the missionaries accompanied them on the journey, including Cyrus Byington and Alfred Wright (Debo, 1934). Almost immediately after arriving in the new Choctaw Nation, they began to establish churches, the first of which was Wheelock Church in 1832 with 37 members (Debo, 1934). Wheelock continues to hold a central place in the identity of the Choctaw Nation of Oklahoma today and the church built in the 1840s is the oldest extant church in Oklahoma (Fig. 2).

By 1861, the Choctaw population had decreased to around 12,500, a decline of almost 30%, and the proportion of Christians had grown. Byington claimed that at this point the Choctaw Nation was by all definitions a Christian nation, despite only 20–25% identifying as Christian (Debo, 1934; Kidwell, 1999). The leadership of the Nation, having attended missionary schools, was disproportionately represented in this 20–25% but their influence encouraged others to convert. Choctaw Reverend Allen Wright, for example, was an ordained Presbyterian minister who served as Chief of the Choctaw Republic from 1866 to 1870 (and proposed the name Oklahoma) when Tribal Council meetings started beginning and ending with prayer, an ongoing practice. Today, the Choctaw Nation is officially a Christian nation. The capital, Durant, is named for an early Choctaw Presbyterian minister (Hamill, 2006) and the Nation has adopted the motto "Faith, Family, and Culture." The primacy of faith in this tripartite-identity statement underscores the prominence of Christianity in the lives and identities of many Choctaws. Yet despite sharing this Christian identity with their neighbors, they remain parallel: "Indian churches and congregations are distinct from their white-world counterparts [and] these distinctions form an important part of the symbolic boundaries between white and Indian ethnicity" (Hamill, 2006).

Fig. 2 Wheelock Church, completed in 1846, is the oldest extant church in Oklahoma and home of the first mission established in the new Choctaw Nation after removal in 1832

Traditional Music and Hymns

A 2014 article on hymn-singing in the Choctaw Nation's newspaper, *Biskinik*, mentions that "one of the most distinctive elements of Choctaw Christian church services are the beautiful, powerful hymns sung in the Choctaw language. For many people, Choctaw hymns like #48, and #21 have become as much a part of their Choctaw identity as *bvnaha* bread and stickball. For some, these hymns convey a relationship with God that defines them as Choctaw people." In other words, Christianity, and hymn-singing in particular, are central components of Choctaw identity for many Choctaw people today.

Choctaw music prior to the introduction of Christian hymns included songs sung both in the Choctaw language and using chanted syllables called vocables. While many traditional songs are no longer known, many are still sung both for Choctaw audiences and at other public performances. These songs and their corresponding dances are usually referred to by the catchall name "social dance" which suggests their function in strengthening community ties, both internally and externally. While many performers and members of the community, including Dan Isaac of the Mystic Wind Social Dance group (pers. comm), consider these songs sacred, they are not generally considered part of a religious tradition today. By the 1950s, these dances were virtually no longer practiced in Oklahoma, largely due to disapproval from church officials (Howard & Levine, 1990; Wright, 1951). Howard and Levine (1990: 13) mention that missionaries and Choctaw ministers considered these dances the "work of the devil," and "evidence of backsliding." This condemnation of dancing was not specific to Choctaws though, and reflected broader opposition to dancing from some Presbyterian churches since at least the early nineteenth century (Rice, 1847).

In the 1970s, with a growing interest in revitalizing Indigenous traditions, Choctaw groups from Oklahoma began traveling to Mississippi to relearn the social dances, paradoxically often under the organization of church groups (Howard & Levine, 1990; Kirven, 2023; Wright, 1951). Gene Wilson, a Choctaw Presbyterian minister, led many of these trips, which were partially funded by a grant from the Presbyterian church (Kirven, 2023). According to Kirven (2023), Curtis Billy, who participated in some of those trips, said "We knew we were Choctaws; we have a language, we did church activities, but … we didn't have any dances or those types of things." Here, when Billy describes how he and others knew they were Choctaw, he lists language and church activities as the two defining features of that cultural identity. Despite many Choctaws never having heard traditional songs or being aware of dances, they still had "a strong cultural identity in part because of their participation in all-Indian gospel-singing groups… as a means of expressing their American Indian identity" (Pesantubbee, 1994: 96). Pesantubbee attributes this to the use of the Choctaw language in churches and emphasizes that "Choctaws retained their cultural identity within the early mission churches" (p. 96). Choctaw hymns were a central element in maintaining Choctaw identity during the transition from traditional religion to Christianity. A defining characteristic of Choctaw ceremonies was "its songs, which were forms of prayer" (Pesantubbee, 1994: 98–99). Traditional ceremonies conducted by an *alikchi* "doctor" around marriage, hunting, and stickball, often involved singing, and later, the "same expressions found their way into gospel songs, especially those written by Choctaws" (Pesantubbee, 1994: 99).

Missionary Cyrus Byington reportedly wrote the first Christian hymn in the Choctaw language in 1824, with nine more hymns following in the subsequent year by unknown authors (Biskinik, 2014). By 1829, the ABCFM were preparing to publish 60 hymns in a first edition of the hymnal, the *Ʊba Isht Taloa Holisso* (ABCFM, 1829a: 365). The sixth edition, published in 1872, has long been considered the standard, and newer versions, including the 1995 edition (2023 reprint) published by the Choctaw Bible Translation Committee and the 2017 version published by the Global Bible Society (GBS), are based on the 1872 edition. While these hymns are often referred to as translations, nearly all are original textual compositions in the Choctaw language. Of the known composers of the 168 canonical Choctaw hymn texts, at least 75 were written by non-Choctaw authors, 29 definitely by Choctaw authors, 8 by authors for whom only initials are known but are presumed to be Choctaw, 55 texts by unknown authors also thought to be Choctaw, and one collaboration between a Choctaw author and a missionary, meaning 93 hymns (55% of those included in the 6th edition) were written or most likely written by Choctaw authors. The hymns are usually referred to by their numbers, or occasionally first lines. The 6th edition included 168 hymns, but more have been added in recent years, and one *Yakni Shoh* (#171) has been commonly sung for decades but was only added to the hymnal in 1995. The most prolific of the known authors is Loring Williams, with 29 hymns, followed by Alfred Wright with 23. Both were non-Choctaw missionaries. Joseph Dukes was the most prolific known Choctaw author, having written nine hymns, with Peter Pitchlynn and Pliney Fisk tied for second with five each. Only one hymn author is referred to by his Choctaw name, Tʋnʋp In Chʋffa, which means "One Opposer" (GBS, 2017). The story of the composition of this hymn (#70) was included in a letter written by Loring Williams to the *Missionary Herald* dated April 11, 1829, where he mentions "I was surprised, inasmuch as the hymn (which was long, and excellent for sentiment and expression), the metre, and the tune, *altogether originated with this aged Choctaw, who can neither read nor write a letter*" (ABCFM, 1829b: 280, emphasis in the original). Unfortunately, the original melody to this tune seems not to have survived.

In general, the hymns' melodies are part of the Sacred Harp tradition, but many are original, likely composed by Choctaw singers. In performance, the songs are often slow and full of feeling and usually sung *a cappella*. There are exceptions, though, including Hymn #46 as recorded by Clelland Billy and others from the Mary Lee Clark United Methodist Church with piano accompaniment, and various versions of Hymn #11. Stevenson, in his 1977 dissertation analysis of Choctaw hymns, provides this summary: "The appetite of the rural full-blood congregations, however, generally includes only those hymns which are either fast and rhythmical or slow and sustained. To put it simply, if a tune speaks to the feet as well as the heart, it is probably vintage Stamps-Baxter; if it is slow, 'it had better be Choctaw'" (Stevenson, 1977: 179). The slow melodies sung to many of the most popular Choctaw hymns are therefore likely of Choctaw origin, or at least interpretation, while faster melodies originate from Anglo-American traditions. The two melodies for hymn #48, one of which is traditional, and the other, the melody for Amazing Grace, demonstrates this contrast.

Many of the hymns originate from the time of removal, especially several of the most popular ones today. These include hymns #53, #112, and #121, among others. The headings for Hymns #53 and #112 include the phrase "Removal Song" (GBS), as well as the description "Meditation on Death" for Hymn #112. Hymn #121 is also associated with removal, according to elder Lillie Roberts, a long-time language teacher and member of the *Chahta Anumpa Aiikhʋna* singing group who has recorded the hymn (pers. comm.). This period was especially poignant for development of the hymns for several reasons. First, the "Great Awakening" in the Choctaw Nation started in 1828, right before removal. Second, the first edition of the hymnal was printed the following year, 1829, meaning that the hymns represented a new cultural innovation, something to distract from and commemorate the devastation accompanying removal. Reports from the time describe the Christian Choctaws as "some of the most religious people" one boat captain had seen, as well as that "their praying and singing made the passage seem like a continued meeting" (Morrison, 1932: 61). One report goes on to say that removing these Christian Choctaws was half the trouble of removing the traditional ones, emphasizing the benefit of

Christian conversion to the colonial project. Today, some elders, including Paula Carney, carry on the oral tradition that the sound of Choctaw singing along the Trail of Tears could be "heard echoing through the trees" (Youngman, 2024). The hymns' connection to the removal era underscores their role in connecting Choctaw people today to their ancestors, as well as to a shared identity shaped by this era. After all, the Choctaw Nation of Oklahoma would still be in the Mississippi homeland it not for removal. Choctaw Christianity and hymn-singing served as tools to forge a new identity in a new land from the Indigenous traditions and outside influences they brought with them, while maintaining connections with the past.

Choctaw hymns are not the only creolized indigenous hymn tradition in Oklahoma. Sterlin Harjo's film, *This May Be the Last Time,* documents Muscogee and Seminole hymns and hymn-singings, and in a subsequent commentary, researcher Hugh Foley describes the creolized origins of the hymns as stemming from the pre-removal era and combining African spirituals, traditional Muscogee songs, Scottish Christianity and singing traditions, all in the Muscogee language (KGOU Radio, 2014). The influence of African American musical traditions is undoubtably present in Choctaw hymns as well, and several popular Gospel songs, including *Swing Low, Sweet Chariot* are credited to Wallace and Minerva Willis, Choctaw "Freedmen" from Oklahoma (Michener, 2010). For many Kiowa Christians, hymns and Christianity are only the most recent iteration of their long-standing, traditional musical and ceremonial relationship with the Creator. According to noted Kiowa hymn singer Ralph Kotay, the main difference between Kiowa traditional religion and Christianity is the addition of Jesus (Lassiter, 2001). For the Kiowa community, all religious songs can signify a relationship with "spirit" and the divine, and the language used in songs to refer to God prior to and after the missionaries' arrival remains the same (Lassiter, 2001). In fact, the use of the Kiowa language is an integral part of their specific practice of Christianity since many elders view the language as a gift "God gave to them specifically and uniquely" (Lassiter, 2001: 346).

Language of the Hymns

When missionaries translated Christian concepts into the Choctaw language, their strategy included taking existing terms in the Choctaw language and adapting them to biblical concepts. This approach represents a type of semantic shift, where existing terms take on new meanings. One example is the term *kashofa*, which Byington et al. (1915) defines as "clean," "purgated," "purified," and "forgiven" among other definitions. Cleanliness, or *kashofa*, had an important role in traditional ceremonial life. Adair (1986 [1775]: 126) dedicates three chapters to describing traditions of cleanliness and purity among Southeastern Native communities. He starts with the tradition of the entire population bathing in rivers every morning, even if the water is frozen: "After bathing, they return home, rejoicing for having so well performed their religious duty, and thus purged away the impurities of the previous day." While he does not mention the word *kashofa* specifically, this was almost certainly the term used. It is unknown whether this term had connotations of forgiveness prior to Christian missionaries' arrival, but eventually they decided this was the best analogue for this central Christian tenet. The word, or its form *kashoffi*, is used with this sense frequently in at least 20 of the hymns, written by both Choctaw and non-Choctaw authors, including the one written by Tʋnʋp In Chʋffa (#70, Fig. 4) and some of the most popular ones today including #48 and #53. We are left wondering whether early Choctaw Christians conceptualized purification in the new religion as distinct from traditional forms or, as seems more likely, a continuation of them.

Other terms underwent similar semantic shift. As previously mentioned, the term *ʋba* was used in traditional Choctaw society to mean both "above" as a literal direction, and also in reference to the deity *Hʋshtahli*, both by itself, as was the case for the Bayou Lacombe Choctaws in 1909 as well as in the form *Ʋba Pi̱ki* as described by missionaries (Bushnell, 1909; Wright, 1828). The association with the traditional Southeastern deity did not dissuade the missionaries from adapting both terms to Christian concepts. In the hymns, the phrase *Ʋba Pi̱ki* is used often, as is *ʋba* by itself, and in other constructions including in the names of the hymns themselves, *Ʋba Isht Taloa*, meaning "to sing above with," and the hymnal itself *Chahta Ʋba Isht Taloa Holisso* or "the Choctaw book to sing above with." In Hymn #5, for example, missionary Loring Williams uses the phrase *Ʋba Pi̱ki*, although he was likely aware of its connotations in traditional religion. Regardless of the missionaries' intended meaning, this term connects back to the traditional theology of both the Choctaw people and their Mississippian ancestors, who conceived of an afterlife in a celestial land above. In addition to *shilup i̱ yakni*, "the spirits' land," this place was likely referred to by its location: *ʋba yakni* or "land above." This term appears frequently in the hymns, sometimes modified slightly, including in #46, written by Choctaw leader Peter Pitchlynn (see Fig. 3). This phrase and its variations appear in 33 out of the 168 canonical hymns (almost 20%). Hymn #159 uses the term *hefen*, which is a borrowing of "heaven," but includes a footnote that clarifies that the term means *ʋba yakni* (land above). Similarly, the term *yakni achukma*, "good land," appears in at least 15 hymns, occasionally with *ʋba* as a modifier, including in hymn #11. The idea of *yakni achukma* was so important for missionaries and Choctaws that one of the first missions and a school in Oklahoma were given that name, later called Goodland in English. It is difficult to imagine that conversion to Christianity removed all traditional conceptions of the afterlife in the "good land above" that Wright and Adair had described. It seems more likely that these terms meant slightly different things to the Choctaws than they did to the missionaries.

One of the clearest examples of the creolized tradition also appears in hymn #46 mentioned above. Verse 3 says "*Hopaii ʋlheha moma kʋt ilʋppak a̱ maha̱ya tok. A̱ yoshoba i̱ filʋmmit ai ahli bat hieli tok.*" An approximate translation of this is "All of the prophets continued here long ago. They stood in the truth, apart from the place of the lost." The use of the term *hopaii* here is a direct connection to pre-Christian prophets. Cushman (1899) describes a *hopaii* (which he defines as "seer") using traditional ceremonies to decide whether to join Tecumseh's efforts to push out the Americans, recounting a story told to him by the author of this hymn's father, Col.

ƲBA ISHT TALOA 46. L. M.

Jesus my all to heaven is gone.

1 An Chisʋs ʋt ʋba yakni
Talaia ka ia hatok:
Anukcheto li hokʋto
Yʋmmak o̱ a̱ biekʋshke.

2 I̱ hina hanta pulla ka
Foyukʋt maha̱ya li hosh,
Chisʋs a binili hoka
Ona likmʋt, pisa la he.

3 Hopaii ʋhleha moma kʋt
Ilʋppak a̱ maha̱ya tok.
A yoshoba i̱ filʋmmit,
Ai a̱hli bat hieli tok.

4 Mi̱ko i̱m ata̱ya yokʋt
Aholitopa bieka,
Ai iti̱ hullo hina ya̱
Fohkʋt a̱ya li pullashke.

Hymn 46 L.M.

1. My Jesus is long gone to that land above.
   I put my trust in him, it shall always be there.

2. I am continuing to follow his bright path.
   When I arrive, I will see where Jesus sits.

3. All the prophets continued long ago here.
   They turned away from the lost place and stood in the truth.

4. The Chief's new way is holy. I will continue to travel the path of care for each other.

Fig. 3 Verses 1–4 of Hymn #46 written by Peter Pitchlynn from the 1872 sixth edition of Wright & Byington's *Chahta Ʋba Isht Taloa Holisso*, or *Choctaw Hymnal*. Notice the use of the terms *ʋba yakni* (land above), *hina hanta* (bright path), and *hopaii* (seer) (listen at https://www.youtube.com/watch?v=W94YInlrhWY as of October 11, 2025)

John Pitchlynn (Cushman, 1899: 320). In this verse, Peter Pitchlynn is using the distant past marker *tok*, suggesting that he is referring to Choctaw *hopaii* in addition to, or perhaps excluding, biblical ones. He also uses the name *Chisʋs*, "Jesus," in the first two verses, however, showing that this is indeed a Christian hymn.

A final example of traditional religious terminology in hymns is *hina hanta* or the "bright path," described above. In the hymns, the phrase *hina hanta* appears regularly. It is found in at least 16 hymns, along with variations in two others, namely *yakni holitopa hina*, meaning "blessed land path," and *hina malʋtha*, "shining path." These usages strengthen the connection to older religious concepts and maintain consistency with contemporary Christian theology emphasizing the unbroken tradition from pre-Christian religion through to today, as the *Hina Hanta* website says (Choctaw Nation of Oklahoma, 2025). Taken together, traditional religious terms are found in nearly all the most popular hymns, including those written by both Choctaw and non-Choctaw authors. This linguistic evidence connects these Christian hymns with pre-Christian Choctaw theology supporting the claim that Choctaw Christianity represents a creolized religious tradition.

Conclusion

Choctaw Christianity and the hymns sung in Choctaw churches make these churches unique among the broader Christian community. Choctaw Christianity originates in the tumult of the 1820s and 1830s, during a time of deterritorialization when

Ʋba Isht Taloa 70. C. M.	**Hymn 70**
1 Himak nitak fehna hoka Pim anukfila ha̱ Ish pi̱ kashoffi pullashke; Chisʋs pi̱ Shahli ma!	1 This very day you surely cleanse our thoughts; O, Jesus is greater than us!
2 Ʋba Pi̱ki Chihowa hʋt Ish apihisa hosh, Ik yʋmohmi, ish ahni na, Hatak e toba tok.	2 Jehovah our father above, as you watch over us, Let it be as you wished when we became men.
3 Hatak ʋt okcha̱ya chi̱ ho̱ Chihowa Ushi mʋt Im anumpa holitopa Hatak a̱ ima tok.	3 So that man would live, That son of Jehovah gave his holy word to him.
4 Chisʋs pi̱ Shahli chia ma! Chu̱kʋsh yohbi fehna Auet ish pima pullashke, E chim ilbʋshʋshke.	4 Jesus you are greater than us! You indeed gave us a serene heart, so we shall pray to you.
5 Himak nitak fehna hoka, Ʋba anumpa ha̱ Haklo pi bʋnna fehna hosh, Il itʋnahʋshke.	5 Because this very day we want to hear the word above, We shall gather together.
6 Pit e chim anumpohonli Auet ish haklashke; Shilombish Holitopa ya̱ Ish pima pullashke. Tʋnʋp in Chʋfa.	6 Just as we are speaking to you, you shall listen to us You give us the Holy Spirit. Tʋnʋp In Chʋfa.

Fig. 4 Hymn #70 from the 1872 sixth edition of Wright & Byington's *Chahta Ʋba Isht Taloa Holisso*, or *Choctaw Hymnal*. This is the only hymn attributed to a Choctaw author using their traditional Choctaw name, Tʋnʋp In Chʋffa. Notice the use of the terms *kashoffi* (forgiveness) and *Ʋba P̱iki* (our father above). Translation by the present author.

Choctaw leaders were adopting and adapting Christianity. For many Choctaws, the shift from traditional religion to Christianity represented a merger rather than a wholesale rejection of one for the other. Instead, a creolized belief system emerged, and with it a new identity and art form. The development of Choctaw Christianity was an effort to preserve a distinct identity as Choctaw people, despite removal from their homeland and the expanding Anglo-American cultural sphere. It was an act of resistance to cultural and community erasure. The preservation and negotiation of this new identity became strongly connected to removal and the establishment of a new Nation in a new land, as did the Choctaw hymns. And while the majority of tribal citizens did not convert, at that time nor until much later, many, including the leadership, did. Still today, despite having declared itself an officially Christian Nation, many Choctaw citizens do not identify as Christian and there is a growing traditionalist movement seeking to recover and reconnect to older cultural practices. Yet, for many, Choctaw identity is deeply tied to Christianity, partially because of its unique connection to older cultural practices. As Kidwell (1999: 151) says, "Culture changed, but even the white man's institutions took on peculiarly Choctaw character."

The specific hymn singing tradition in these churches makes them unique from the universe of Protestantism found nearby in Oklahoma and the United States as a whole. This hymnody is connected to traditional religious beliefs, as can be seen through the presence of pre-Christian religious terminology in the hymns. Participating in hymn-singings is part of singers' identities as Choctaw people, their connections to their ancestors and to each other. Even though many of the hymns are sung to Christian tunes, they are distinctly Choctaw, both in origin and in practice, and they represent the creolized belief system of Choctaw Christianity. While this hymnody shares similarities with other Indigenous, African American, and Anglo-Southern hymn-singing traditions, it remains a unique tradition defined by the use of the Choctaw language and the shared experiences and beliefs of Choctaw people.

These hymns still have much to tell us, including about how communities can use musical traditions for language revitalization, and how this specific musical tradition has preserved a highly formalized variety of the Choctaw language. With more attention, there is the possibility that the popularity of hymn-singings will increase, leading more community members, especially young people, to connect to their Choctaw identities using this unique art form, something that is happening already in a limited way (Wallace, 2023). A systematic analysis of original hymn melodies as well as a detailed comparison of hymns and social dance traditions expanding on Draper (1982) would answer important additional questions about the musical elements of this creolized tradition.

Acknowledgements Research for this chapter was made possible in part by a residency at A Studio in the Woods.

References

ABCFM (American Board of Commissioners of Foreign Missions). (1829a). Hymns in the Choctaw language. *Missionary Herald, 25*.

ABCFM (American Board of Commissioners of Foreign Missions). (1829b). Journal of Mr. Williams at Ai-Ik-Hun-Na. *Missionary Herald, 25*.

Adair, J. (1986 [1775]). *History of the American Indians*. Promontory Press.

Biskinik. (2014, February). *Chahta Vba Isht Taloa: The Choctaw Hymns. Iti Fabvssa.*

Bushnell, D. I. (1909). *The Choctaw of Bayou Lacomb, St. Tammany Parish, Louisiana*. American Bureau of Ethnology, Bulletin 48. Government Printing Office.

Byington, C., Swanton, J. R., & Halbert, H. S. (1915). *A dictionary of the Choctaw language*. Bureau of American Ethnology, Bulletin 46. Government Printing Office.

Carson, J. T. (2003). *Searching for the bright path: The Mississippi Choctaws from prehistory to removal*. University of Nebraska Press.

Choctaw Bible Translation Committee. (2023 [1995]). *Chahta Ʊba Isht Taloa Holisso*. Choctaw Bible Translation Center.

Choctaw Nation of Oklahoma. (2025). *Hina Hanta | The bright path*. Retrieved July 14, 2025, from https://hinahanta.choctawnation.com/ProficioWebModule/
Cushman, C. (1830, March 10). Additions to the church. *Missionary Herald, 26*(5).
Cushman, H. B. (1899). *History of the Choctaw, Chickasaw, and Natchez Indians*. Headlight Printing House.
Debo, A. (1934). *The rise and fall of the Choctaw Republic*. University of Oklahoma Press.
Draper, D. E. (1982). Abba Isht Taluwa: The Christian Hymns of the Mississippi Choctaw. *American Indian Culture and Research Journal, 6*(1) https://doi.org/10.17953
Ellis, E. N. (2022). *The great power of small nations: Indigenous diplomacy in the Gulf South*. University of Pennsylvania Press. https://doi.org/10.9783/9781512823189
Folsom, R. (1830, May). Letter from Captain Robert Folsom, Dated at Hebron, 8th of March, 1830. *Missionary Herald, 26*. American Board of Commissioners of Foreign Missions.
Fundaburk, E. L., & Foreman, M. D. F. (2001). *Sun circles and human hands: The Southeastern Indians art and industries* (1st ed.). University of Alabama Press.
Gatschet, A. S. (1884). *A migration legend of the Creek Indians, with a linguistic, historic, and ethnographic introduction*. D.G. Brinton.
GBS (Global Bible Society). (2017). *Chahta Ʋba Isht Taloa Micha Taloa Ikbi Holisso Hoke | Choctaw Hymn and Tune book*. Global Bible Society.
Hamill, J. (2006). *Going Indian*. University of Illinois Press.
Hiemstra, W. L. (1948). Presbyterian missionaries and mission churches among the Choctaw and Chickasaw Indians, 1832–1865. *Chronicles of Oklahoma, 26*(4), 459–467.
Howard, J. H., & Levine, V. L. (1990). *Choctaw music and dance*. University of Oklahoma Press.
Johnson, P. C. (2002). *Secrets, gossip, and gods: The transformation of Brazilian Candomblé*. Oxford University Press.
KGOU Radio. (2014, November 17). *Muscogee Creek Hymns influenced by congregational line singing?* https://www.kgou.org/arts-and-entertainment/2014-11-17/muscogee-creek-hymns-influenced-by-congregational-line-singing
Kidwell, C. S. (1999). The language of Christian conversion among the Choctaws. *The Journal of Presbyterian History, 77*(3), 143–152.
Kirven, S. (2023, May 1). Remembering the Choctaw cultural revitalization movement. *Biskinik*. Retrieved September 30, 2025, from https://www.choctawnation.com/biskinik/news/remembering-the-choctaw-cultural-revitalization-movement/
Lankford, G. E. (2004). World on a string: Some cosmological components of the Southeastern ceremonial complex. In R. F. Townsend (Ed.), *Hero, hawk, and open hand: American Indian art of the ancient Midwest and South*. Art Institute of Chicago and Yale University Pres.
Lassiter, L. E. (2001). From here on, i will be praying to you: Indian churches, Kiowa Hymns, and Native American Christianity in Southwestern Oklahoma. *Ethnomusicology, 45*(2).
Lindenfeld, D., & Richardson, M. (2012). Introduction. Beyond conversion and syncretism. In D. Lindenfeld & M. Richardson (Eds.), *Beyond conversion and syncretism: Indigenous encounters with Missionary Christianity, 1800-2000*. Berghahn Books.
Michener, J. (2010). Willis, "uncle" Wallace and "aunt" Minerva. In *The encyclopedia of Oklahoma history and culture*. https://www.okhistory.org/publications/enc/entry?entry=WI018
Morrison, W. B. (1932). *The Red Man's trail*. Presbyterian Committee of Publication.
Neylan, S. (2003). *The heavens are changing: Nineteenth-century protestant missions and Tsimshian Christianity*. McGill-Queen's University Press.
Pesantubbee, M. E. (1994). *Culture revitalization and indigenization of churches among the Choctaw of Oklahoma*. University of California.
Rice, N. L. (1847). *A discourse on dancing*. Holland and Hitchler.
Spitzer, N. R. (2003). Monde Créole: The cultural world of French Louisiana Creoles and the creolization of world cultures. *Journal of American Folklore, 116*(459), 57–72.
Stevenson, G. W. (1977). *The hymnody of the Choctaw Indians of Oklahoma*. The Southern Baptist Theological Seminary. Unpublished dissertation.

Swanton, J. (1931). *Source material for the social and ceremonial life of the Choctaw Indians*. Government Printing Office.

Wallace, K. L., Jr. (2023). Chahta Uba Isht Taloa—Songs that make us Choctaw. *The Hymn, 74*(4).

Wright, A. (1828, June). Choctaws: Religious opinions, traditions, etc. *Missionary Herald, 24*. American Board of Commissioners of Foreign Missions.

Wright, A., & Byington, C. (1827). *Chahta Holisso Ai Isht Ia Ummona | The Choctaw spelling book* (2nd ed.). Morgan, Lodge, & Fisher.

Wright, A., & Byington, C. (1873). *Chahta Uba Isht Taloa Holisso* (6th ed.). Presbyterian Committee of Publication.

Wright, M. H. (1951). *A guide to the Indian Tribes of Oklahoma*. University of Oklahoma Press.

Youngman, R. (2024, September 27). Season 7, episode 6 part 2 our Choctaw Hymns and ancestral stories with Paula Carney and Cheryl Weeden, Choctaw. *Native ChocTalk*. https://nativechoctalk.com/season-7/season-7-episode-6-part-2-our-choctaw-hymns-and-ancestral-stories-with-paula-carney-and-cheryl-weeden-choctaw

Parallel Universes: The Racial Divide in the Segregated United States

Allison Schlobohm, Stephanie Mahin, and Lori Boyer

Introduction

In much of segregated America during the mid-twentieth century, Black and White citizens lived in parallel universes. De jure and de facto segregation, including Jim Crow laws, reigned supreme. Separate public facilities, including water fountains and swimming pools; separate school districts; separate neighborhoods; unequal access to public services and amenities; policed boundaries; and general racist animus meant that people of different races living in the same city, town, or county led fundamentally different lives. Figure 1 displays a 1929 "White Only" sign from the St. Louis Southern Railway.

This chapter documents, using interviews and historical research, some of the everyday geographical experiences of Black American citizens who inhabited these parallel universes. The authors employed phenomenological geographical methods through the lens of social geography, taking seriously the concept of spatialization, or the idea that, to understand space, one must understand the ways that humans in their social relationships, movements, and understandings are impacted by and co-create the environments that they exist within (Massey, 1991; Pickles, 1985; Seamon, 2023).

This project is unique in its phenomenological methods and focus on racialized geographical locations within the United States. Most contemporary academic research related to personal experiences of racial isolation is primarily concerned with current conditions and/or on non-U.S. locations (e.g. Catney et al., 2019; Gillette & Hurley, 2018; Hanhorster, 2000; Marsh et al., 2010; Raanan & Shoval, 2014; Smiley, 2013). Contemporary research that does examine historical

A. Schlobohm (✉) · S. Mahin · L. Boyer
University of North Carolina, Chapel Hill, NC, USA
e-mail: Allison_Schlobohm@Kenan-Flagler.unc.edu; Stephanie_Mahin@Kenan-Flagler.unc.edu; Lori_Boyer@Kenan-Flagler.unc.edu

S. D. Brunn (ed.), *Transdisciplinary Research Across Knowledge Worlds*,
https://doi.org/10.1007/978-3-032-19491-6_9

Fig. 1 A 1929 sign from the St. Louis Southern Railway, likely informing southbound passengers that they would be forced into separate cars (Photo courtesy of the Missouri Historical Society; used with permission)

geographies of racial segregation in the United States often employs publicly available data not derived from personal experience or interviews, such as census tract information, political and legal documents, and real estate assessment maps (e.g. Rey & Knaap, 2024; Rothstein, 2017; Trounstine, 2018).

Such explorations of non-U.S. contexts, contemporary conditions, and publicly available historical data are no doubt essential, but they omit a key part of the story: personal experiences of racist division and constrained mobility. To begin to understand the embodied consequences of Jim Crow geographical separation in the mid-twentieth century United States and the social geographies of said separation, one must learn from the Black Americans who experienced them. Unfortunately, these personal histories are at risk of being lost since the citizens who can best remember the United States in the 1940s, 1950s, and 1960s are now in their golden years.

For this chapter, phenomenological interviews were conducted with 11 Black participants regarding 4 U.S. locations segregated during the mid-twentieth century: Tulsa, Oklahoma; Kinloch, Missouri; Birmingham, Alabama; and Edgecombe County, North Carolina. During the interviews, interviewees created maps of their experiences, layering them onto GIS data using "My Maps" in Google Maps. Interviewee ages (in late 2024 and early 2025) ranged from 61 to 99. As a result, the maps created are spatially focused and not temporally bound. Historical research supplemented this focus.

Following the interviews, audio recordings were uploaded, transcribed, and coded using MAXQDA, a data analysis software. Researchers conducted multiple rounds of coding using grounded theory to identify emerging themes within each interview and developed a shared code list. Throughout the process, researchers engaged in memo writing to document analytical highlights, refine coding decisions, and track evolving interpretations. All interviews were then re-coded iteratively until overarching themes emerged at the project level, ensuring both analytical depth and consistency.

The remainder of this chapter is organized by geographic location, with each location organized by theme. One theme, "the racial divide," appears in each section, as it was the most consistent subject across all coded interviews. This theme, in general, references the ways that space, place, and experience were shaped along racial lines within each of the locations. Other themes appear in just one section even though they were relevant to each geographical area: boundaries and Black movement (Tulsa), neighborhood destruction (Kinloch), integrating neighborhoods (Birmingham), and entrepreneurship and economic factors (Edgecombe County). The chapter ends with a discussion of the importance of this research and areas of potential future focus.

Tulsa, Oklahoma: A City of Contrasts

Tulsa, known in the early twentieth century as the "Oil Capital of the World," attracted Black migrants seeking economic prosperity and "relative freedom" (Brophy, 2003). Settling North Tulsa signaled Black progress and racial uplift (Ike, 2024). However, during the Jim Crow era of the 1950s and early 1960s, Tulsa, like many American cities, was marked by racial violence (Sullivan, 2004). Tulsa exhibited stark contrasts—White residents lived freely, while Blacks experienced isolation within their communities.

James O. Goodwin, an 85-year-old life-long Black Tulsan, is an attorney and publisher of *The Oklahoma Eagle*, one of the oldest Black-owned newspapers in the country (The Oklahoma Eagle, 2025). As a research participant, he recalls the separation, "My history is a separate and unequal world. That's where I grew up. That was my whole life. That's the only world I knew … but within that world … was some of the best people in the whole world."

The Goodwin family migrated to Tulsa in 1916, contributing to a thriving Black community with dedicated barbers, beauticians, preachers, teachers, entertainers, attorneys, and entrepreneurs. Goodwin describes,

> … when the Cherokees or the … tribes came from the Southeast, they brought slaves with them. And so when emancipation came, there were those who were able to hold on to their land without it being taken and they were prosperous. And so you had this income in this area, and … reportedly … money was circulated maybe 28 times before [it] left the Black community.

The Racial Divide

The Greenwood District was a residential and commercial region separated from the rest of the city by an imaginary line. However, the area was nestled within North Tulsa, a larger geographic area of neighborhoods, schools, and churches, which was heavily populated by the Black Tulsan community. Attorney Goodwin described this divide as "Black immunization" rather than isolation, a protective measure against racial violence. Black residents established roughly 644 blocks of self-contained community space (Schneider, 2002) (see Fig. 2).

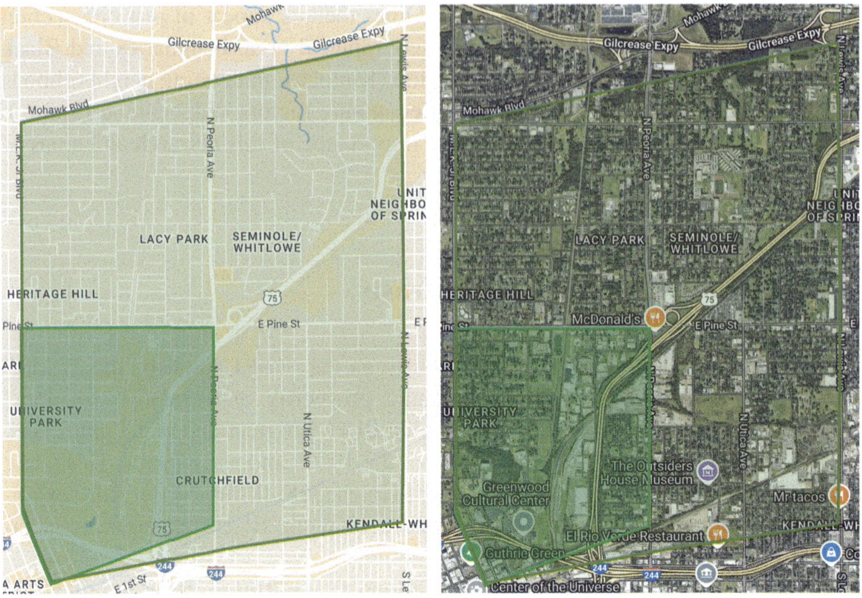

Fig. 2 Racial divide: The smaller green area is the 644-square block area in North Tulsa where Blacks felt free to move as remembered by Attorney Goodwin. The larger areas in green are the boundaried areas as remembered by Dr. Ford Wimberly (Google. (n.d.). Tulsa Safety Map)

Dr. La Verne Ford Wimberly, a retired Black educator for Tulsa Public Schools, and another research participant, noted:

> … we just kind of stayed in our own community. Plus, we were able to see our teachers at school, see them in the church, see them in the doctor's office, the grocery store, wherever you went. You just saw the community, and everybody took hold of you at that point.

Like other segregated cities during the Jim Crow era, Tulsa erected physical "signposts" or barriers of racial demarcation—or spatial strategies of white supremacy as Weyeneth (2005) calls them—that enforced segregation across historically public spaces. Commonly seen were "whites only" or "coloreds only" signs at water fountains, restrooms, or on public transportation (Weyeneth, 2005). However, in many areas, particularly residential neighborhoods, racial separation was enforced through explicit legal measures and implicitly maintained through social practices. According to Goodwin and other residents, for Black Tulsans living in the northern part of the city, the racial divide was clearly understood. Ford Wimberly recalls,

> We were confined mostly from the area of Archer to maybe Apache and from going east and west from to about Lewis, and then … west on Cincinnati. And that was basically all we knew for the most part in the early 50s, the late 40s … The only time we really left the neighborhood was perhaps to go downtown, which we could shop at only a couple of stores like Froug's [Department Store] and JCPenney's and something like that.

Over the years, these unspoken boundaries were both established and actively reinforced by Black residents themselves through word-of-mouth. Generations passed down community knowledge about places that were safe and places that were off-limits. This internal enforcement was a protective strategy against racial violence and discrimination. Such self-policing of community boundaries illustrates how racial segregation was not only imposed externally but also strategically upheld internally as a means of survival and protection. Goodwin explains,

> The impact of the racial divide was not evident to me. I do remember vaguely that we were admonished by, I don't know who in the family, not to cross certain streets at certain times of night … that really did not resonate with me because everything we needed was within [our community] and there was no need to go [out of our community].

During the 1921 Tulsa Race Massacre, a White mob destroyed Greenwood, a thriving economic hub with over 200 Black-owned businesses. The violence was sparked by the arrest of a 19-year-old Black teen accused of assaulting a White woman. The allegation was amplified by an inflammatory newspaper headline that incited White mobs to gather and attack the Black community and burn it down (Ike, 2024; Messer et al., 2010). Despite widespread devastation, residents rebuilt Greenwood, reclaiming economic independence and community identity (Ike, 2024; Moreno, 2021).

Additionally, policies like the Federal-Aid Highway Acts of the 1960s deepened Tulsa's racial divide. Interstate 244, ironically named "Martin Luther King Expressway," ran through Greenwood, displacing residents, and businesses, symbolizing modern racial separation (Ike, 2024; Moreno, 2021; Rothstein, 2017) (see Figs. 3 and 4).

Boundaries and Black Movement

Black movement within the city was largely constrained, with North Tulsa being one of few spaces where Black residents could navigate freely. Wheeler (1976) noted that an inability to move about limited economic opportunities like finding jobs. This geographic and social confinement was not merely the result of long-standing social structures. It was actively reinforced by decades of discriminatory policies, redlining, and racially restrictive covenants, all of which systematically shaped Tulsa's urban landscape. Further, any pre-existing landmarks that helped to define Greenwood were destroyed in the 1921 Tulsa Race Massacre (see Fig. 5). The destruction disrupted not only physical structures but also the district's sense of place and community. Despite these disruptions, residents still found ways to rebuild community ties. Ford Wimberley recalled the warmth and connectedness of daily life after Greenwood's reconstruction:

> [We] walked everywhere. We had a couple of grocery stores around the corner from me and you'd walk to the grocery store there, McCree's grocery and Burns Grocery and you'd meet people and talk … it [was] just wonderful.

Meanwhile, Coleman (1999) suggests that dismantling legally enforced racial separation increased Black mobility in some parts of the country, though Tulsa's experience was notably different. Despite landmark rulings like Brown v. Board of

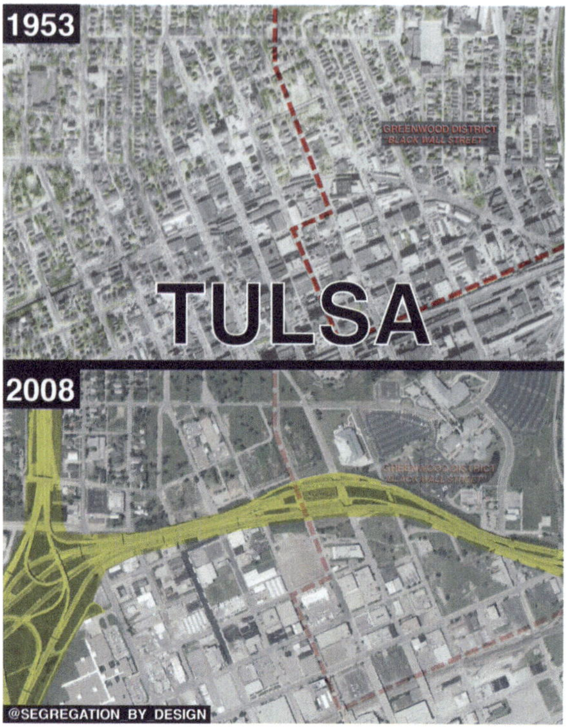

Fig. 3 The famous "Black Wall Street"—before and after highway construction and "urban renewal" (*Instagram* by @ Segregation_by_design, August 1, 2023, https://www.instagram.com/p/CvZ2tDJshMX/; used with permission)

Education (1954) and legislation such as the Civil Rights Act of 1964, Tulsa's racial divide persisted. Legal desegregation alone was insufficient to dismantle the entrenched economic and infrastructural barriers that continued to restrict Black mobility. Although the explicit boundaries limiting where Black residents could move within the city—whether physical or socially constructed—expanded, spatial isolation remained deeply embedded. Ultimately, while desegregation allowed greater freedom of movement, it also exposed stark inequalities between Black and White communities. Systemic policies, including highway construction, discriminatory zoning, and economic neglect, continued to confine Black Tulsans to under-resourced areas, restricting their full participation in the city's economic and social life. Ford Wimberly recalls how she never considered legally enforced racial separation until her high school years.

> … around 1954, 1955 was when they stopped having you to attend your school because of race. And the boundaries were changed, and it was called the neighborhood concept … between my sophomore … and junior year[s][of high school] was when I really thought about race and how we were made to stay in one area and there were some Black students who lived in what we call South Haven, just west of Tulsa … They had to [pass] all these other high schools to get to Booker T. [Washington High School], and then I really started to realize that there was a distinct difference in how we were treated and not treated.

Navigating these shifting and often undefined boundaries, Black North Tulsans continually asserted their presence and agency. Despite enduring spatial isolation and

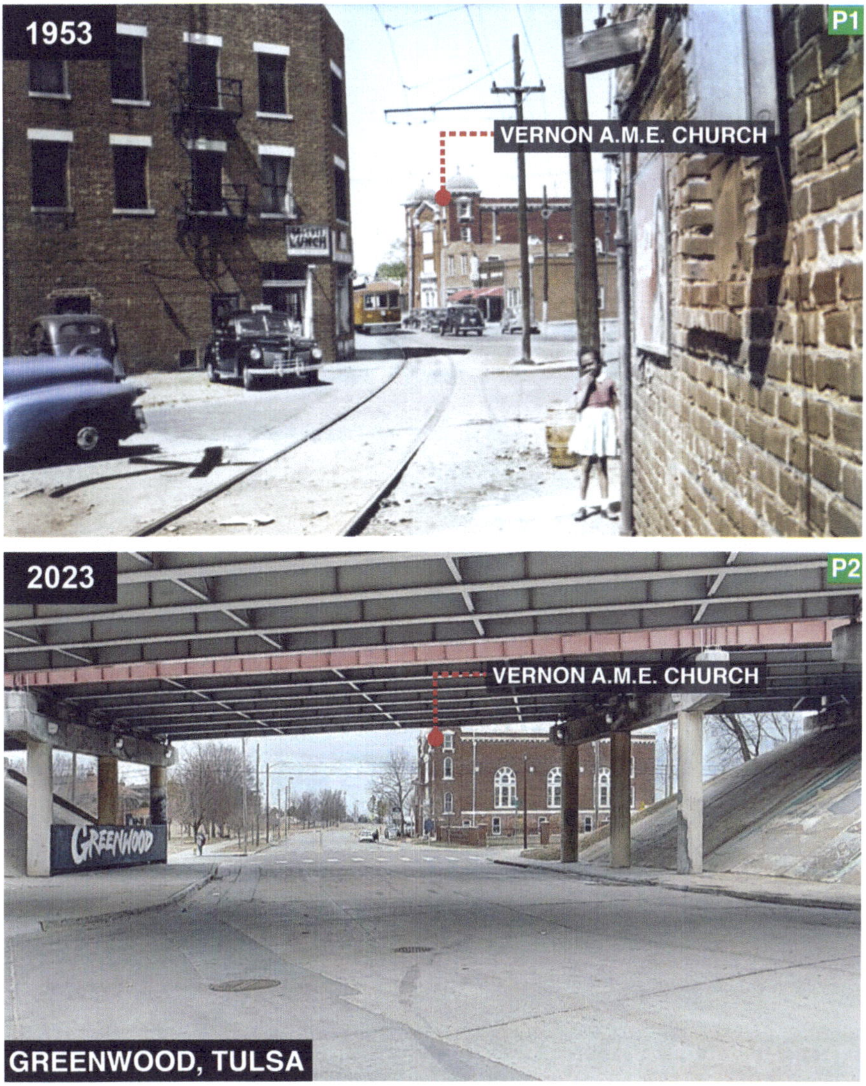

Fig. 4 Interstate 244 displaced residents and businesses in Greenwood (*Instagram* by @ Segregation_by_design, September 22, 2023, https://www.instagram.com/p/CxfslOeMbEU/; used with permission)

Fig. 5 Photos of Greenwood, Tulsa, before and after the 1921 Race Massacre (*Instagram* by @ Segregation_by_design, August 15, 2023, https://www.instagram.com/p/Cv92QE6MwnF/; used with permission)

systemic restrictions, they actively reclaimed economic autonomy and sustained vibrant community networks. Their experiences not only underscore profound resilience in the face of structural racism but also demonstrate Black excellence—reflecting a collective commitment to growth, innovation, and community empowerment within constraints imposed by a deeply divided city.

Kinloch, Missouri: Boxed In and Bought Out

Kinloch, Missouri, was developed in the late nineteenth century as "Kinloch Park," an unincorporated commuter suburb in St. Louis County. At first, a small portion of Kinloch was designated for Black citizens working in neighboring White homes. However, according to the City of Kinloch's history website, when a White landowner sold a parcel in a previously all-White section to a Black citizen, a White exodus ensued and more Black citizens arrived (https://kinlochmo.org/resources/history.php, accessed April 6, 2025). When Kinloch officially incorporated in 1948, it was the first all-Black city in the state of Missouri.

As Kinloch grew and flourished in the decades that followed, its residents were increasingly separated from and disrespected by their White neighbors. From new municipal boundaries to police surveillance, the story of Kinloch is, in part, one of racist division and animus. Another part of Kinloch's story is its unfortunate demise following an airport expansion project in the 1980s and 1990s, during which the City of St. Louis purchased most of Kinloch's homes (Stelzer, 2000). As research participant Mrs. Wilatrel Rice, age 75, put it: "They've been wanting to get it [Kinloch] for a very long time. So, they kept pinching off of it and chopping it off until it's what it is today."

The Racial Divide

Kinloch is surrounded on all sides by two historically White municipalities—Berkeley to the north, west, and south and Ferguson to the east (Hemphill, 2020). Much of what is now Berkeley was once part of Kinloch Park. However, in 1939, White residents voted to create a new municipality to subvert racial integration in the Kinloch schools (Wright, 2000). Once established, one of Berkeley's first actions was to form its all-White school district (Ibid.). The city of Ferguson is also its own municipality, one of 88 in St. Louis County. Figure 6 displays Kinloch's official municipal boundaries—Kinloch in red, Berkeley in black, and Ferguson in blue. One interviewee, 81-year-old Mr. David Rice, explained: "They just hemmed us in."

Ferguson has been nationally recognized as a former "sundown town," a place where Black people were encouraged or forced to leave before dark (Teague, 2019). Through generational knowledge and police enforcement, Kinloch's Black

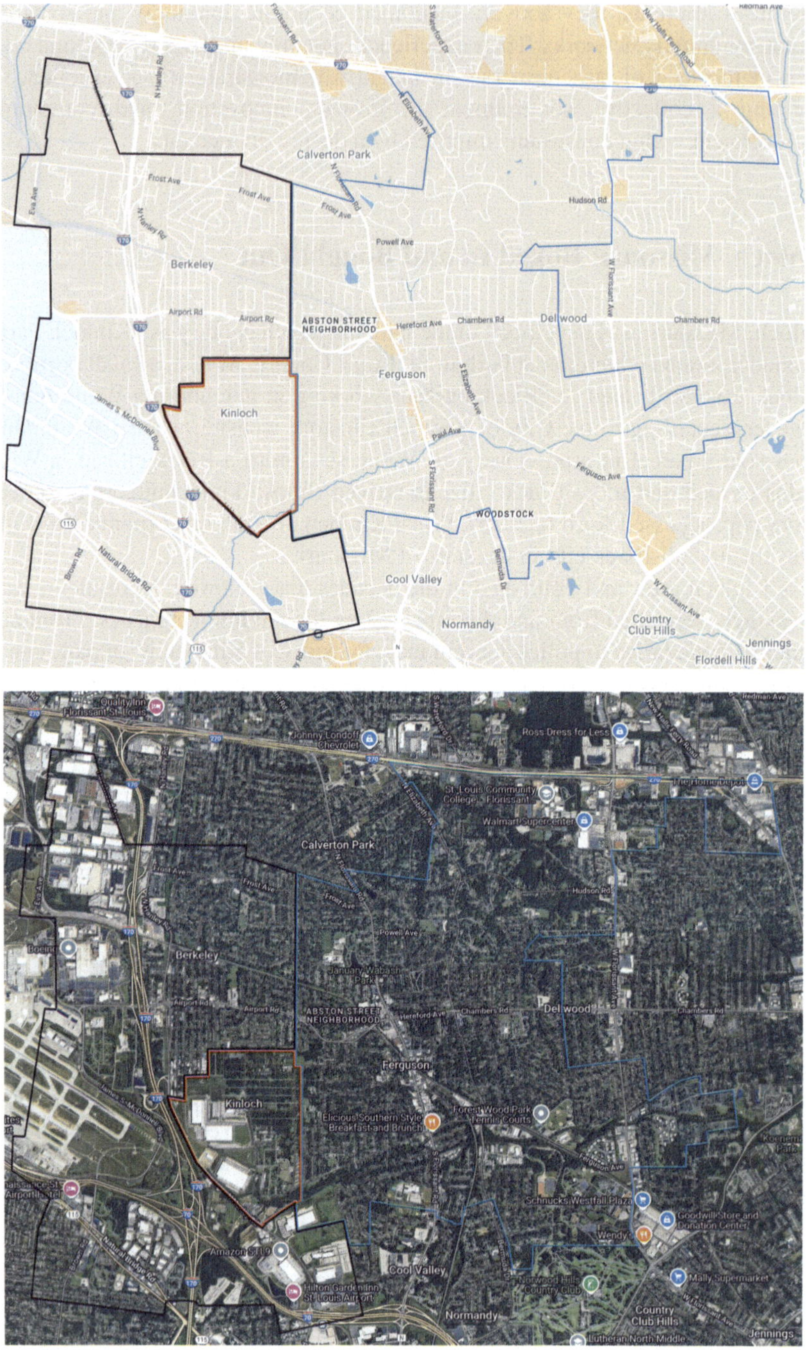

Fig. 6 Municipal boundaries of Berkeley, Kinloch, and Ferguson (Google. (n.d.). Kinloch Maps—Municipal Boundaries)

residents knew where they were and were not allowed to go. According to 86-year-old Mr. James Simmons, a research participant, long-time Kinloch resident, and co-leader of the Kinloch Historical Association, "Ferguson used to turn you around real quick." He and his wife, Mrs. Shirley Simmons, age 85 and also a participant and co-leader of the historical association, described one common encounter:

> Mr. Simmons: There used to be a bridge there [near the intersection of Mabel and Suburban, at the border of Kinloch and Ferguson] but you could only walk across. You couldn't drive across. You could ride your bike across … We used to go over there … and if it was getting dark …
>
> Mrs. Simmons: … you better get away.
>
> Mr. Simmons: They'd see you coming across that bridge; he'd turn you around, say, wait. Come back tomorrow.
>
> Interviewer: Who would turn you around?
>
> Mr. Simmons: Police. Seemed like they'd be there every time…. They'd tell you to turn around, go back. Come back tomorrow. They don't want you over there after dark.

Police patrolled and surveilled Kinloch's boundaries with Berkeley, too. As Mr. Simmons described, "They used to harass you and want to know where you're going."

Figure 7 details Mr. and Mrs. Simmons' map of Kinloch's safe zone (green). Suburban Avenue sits near the southern edge of the map, and the orange line represents the approximate location of the pedestrian bridge.

Although Kinloch citizens often had to leave the safety of their city to find employment, working outside of Kinloch did not give one a free pass to move freely after dark. Mrs. Rice described her mother's work in Ferguson:

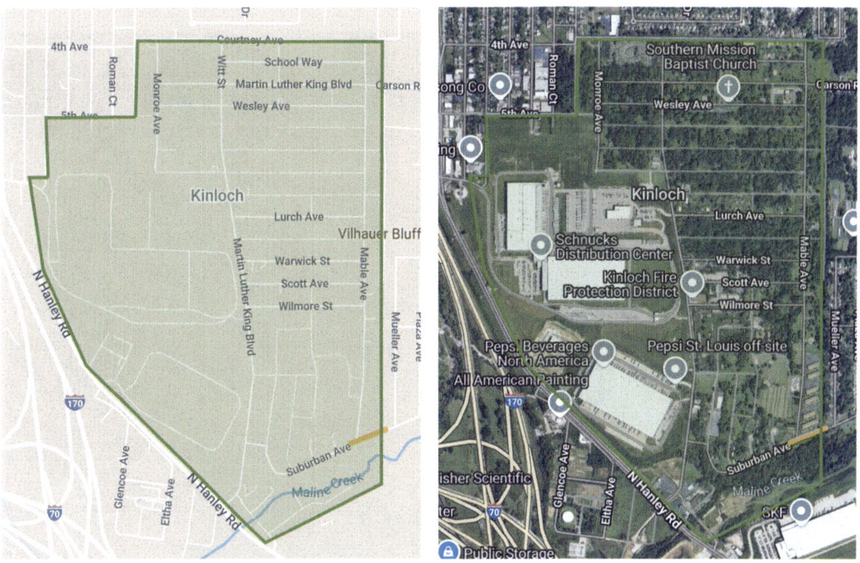

Fig. 7 Mr. and Mrs. Simmons' map of where they felt safe in and near Kinloch (Google (n.d.). Kinloch Maps—Simmons Safety Map)

> My mother did day work when she cleaned houses for White people. And if her ride didn't come before the sun went down, she had to spend the night at the home of the family where she was working … No cab or service car was allowed in the neighborhoods at night.

Figure 8 outlines the map of Kinloch's safety lines that Mrs. Rice constructed alongside her husband, Mr. Rice. The orange line just north of Suburban Avenue

Fig. 8 Mr. and Mrs. Rice's map of where they felt safe and unsafe going in and near Kinloch (Google. (n.d.). Kinloch Maps—Rice Safety Map)

represents a non-street path that Mr. Rice and others would take to get to South Florissant Avenue (in Ferguson) for work.

Though movement in and out of Kinloch was racially restricted, most Black residents were happy to be in their protected enclave. Sadly, this oasis would not last forever.

Neighborhood Destruction

Kinloch was once a shining example of Black community, entrepreneurship, resolve, and excellence, producing such national treasures as Senator Maxine Waters, activist Dick Gregory, and actress/singer Jenifer Lewis (Wright, 2000). As Mrs. Simmons stated when discussing her time in Kinloch, "those were the good old days .... It was just wonderful." Figure 9 provides a glimpse into community life in Kinloch in the 1970s.

Kinloch has been institutionally neglected and exploited for decades, and it now stands in disrepair. At its height in the 1960s, the city of Kinloch had at least 6500 citizens (Missouri Census Data Center, n.d.). In 1980, Kinloch had a population of 4455 (ibid.). In 2020, that number was 263—a 94% population loss since 1980 and 96% loss since 1960 (United States Census Bureau, Kinloch City, Missouri, https://

Fig. 9 Kinloch is often remembered and celebrated as a place of strong and joyful Black community (This picture, provided by documentarian Alana Marie, shows a glimpse into community life in the 1970s; used with permission)

data.census.gov/profile/Kinloch_city,_Missouri?g=160XX00US2938972, accessed April 6, 2025).

The airport buyout program during the 1980s and 1990s is primarily to blame for Kinloch's demise, and neither the Rices nor the Simmonses described the buyout as race-neutral. Mrs. Simmons explained that it was just one piece of a longer attempt to destroy Kinloch and all it represented:

> They've been trying to get Kinloch, and so as time went on, they finally broke us up. And the airport didn't even want it. Because what they said they [were] going to do, they never did do it. So, nothing's out there now but a lot of factories.

Mr. Rice was even more pointed, explaining that the story of Kinloch is one of systemic racism and the city's current state is by design. And, according to him, some of the culprits in Kinloch are the large companies that now sit on land once owned by Black residents. Mr. and Mrs. Rice:

> Mrs. Rice: The roads are completely blocked with mattresses and furnishings and furniture and stuff. And there's at least three churches still there, and you can't get down the road to the church.
>
> Mr. Rice: It's like by design … To me, between the City of St. Louis, Pepsi, Schnucks, St. Louis County, they are allowing this to happen.

Figure 10 shows a historical image of a residential street in Kinloch. Figure 11 is one example of the current state of the city's streets. For the thousands of Black citizens who used to call Kinloch home, and the hundreds that still do, Kinloch's current condition adds insult to injury.

Fig. 10 Historic example of a residential street in Kinloch (Photo provided by Raychel Proudie, published in Schallhorn, 2020; used with permission)

Fig. 11 Current state of Kinloch's streets (Photo provided by Raychel Proudie, published in Schallhorn, 2020; used with permission)

Birmingham, Alabama: Deep Wounds in the Deep South

Birmingham's legacy was forged in the midst of America's Reconstruction period. Founded in 1871—less than a decade after slavery—the new city and the burgeoning country were creating a new society that merged two vastly different experiences. On one path, White residents prospered, enjoying the city's riches and opportunities. On the other, Black residents survived, fighting for financial freedom and physical safety. Both paths converged in a city that was soon to be a Southern industrial powerhouse: Birmingham, Alabama. Birmingham earned the moniker Magic City because of its incredible growth from mining its natural resources. Some referred to Birmingham as the "Pittsburgh of the South" or "Iron City" (Majors, 2022). And later, the city earned other nicknames for its more dangerous interactions between Black and White residents.

As its population grew, Birmingham could no longer quench the fires that ignited its Civil Rights Movement. The city's geography tells the story of racial segregation and managed integration. Two of its residents shared their experiences in Birmingham during interviews for this chapter. Barbara McDonald, 88, is a Black woman who was born, raised, and chose to build a life in Birmingham, accepting but not internalizing the city's trauma. Her niece, LaDonna Wright Cherry, 60, is a Black woman who was born, raised, and chose to leave Birmingham to live her adult life elsewhere.

The Racial Divide

In the early twentieth century, Center Street—which runs west of Interstate 65—separated Birmingham into residential zones, with White residents on the west side and Black residents on the east. Figure 12 shows the 1926 zoning map of Birmingham that segregated neighborhoods by race; areas with cross-hatches were zoned for Black residences.

Black and White residents rarely shared public spaces. Kelly Ingram Park enforced segregated visitation days (see Fig. 13). Birmingham's Lyric Theatre allowed Black and White audiences to watch the same show, though sitting in separate sections (See Fig. 14) (Birmingham Times, 2020). Even the Birmingham Zoo avoided overlapping outings from Black and White elementary school children.

Black parents and community leaders were mindful of the city's implicit and explicit guidelines. They shared their wisdom to keep their communities safe. Mrs. McDonald, who lived in the Smithfield neighborhood along Center Street as a young girl, explained some of the lessons she learned from her parents and others in the community:

> They taught us where we could go, where we couldn't go. They taught us what we could do and what we couldn't do. They taught us that … you must always be aware of your surroundings, and you don't try to move yourself in a space that you know you're not supposed to be in there. You have just as many people here of your own race that you can enjoy. You do not have to get out of your boundaries … Our parents taught us to stay in our lane, and they didn't take us out of our lane.

Figure 15 is this section's research-created map of Birmingham, which outlines two areas considered safe by Black residents. The shape on the left is Mrs. McDonald's memory of the safe zone in the Smithfield community, and the shape on the right is the Fourth Avenue Historic District, where many Black citizens spent recreational time and built businesses.

Fig. 12 Birmingham's 1926 zoning map, which segregated neighborhoods by race. Cross-hatched areas were designated for Black residences (Photo courtesy of the Birmingham Public Library; used with permission)

Fig. 13 During the 1960s, Kelly Ingram Park emerged as a focal point in Birmingham's Civil Rights struggles. The park now features public art installations such as *The Foot Soldier* statue, which symbolizes the confrontation between demonstrators and law enforcement during the Civil Rights Movement (Photo in public domain)

Fig. 14 The Lyric Theatre permitted both Black and White patrons to attend performances, but enforced segregation with designated seating areas within the venue, as demonstrated by this ticket stub (Photo provided by the Lyric Theater, Birmingham Alabama; used with permission)

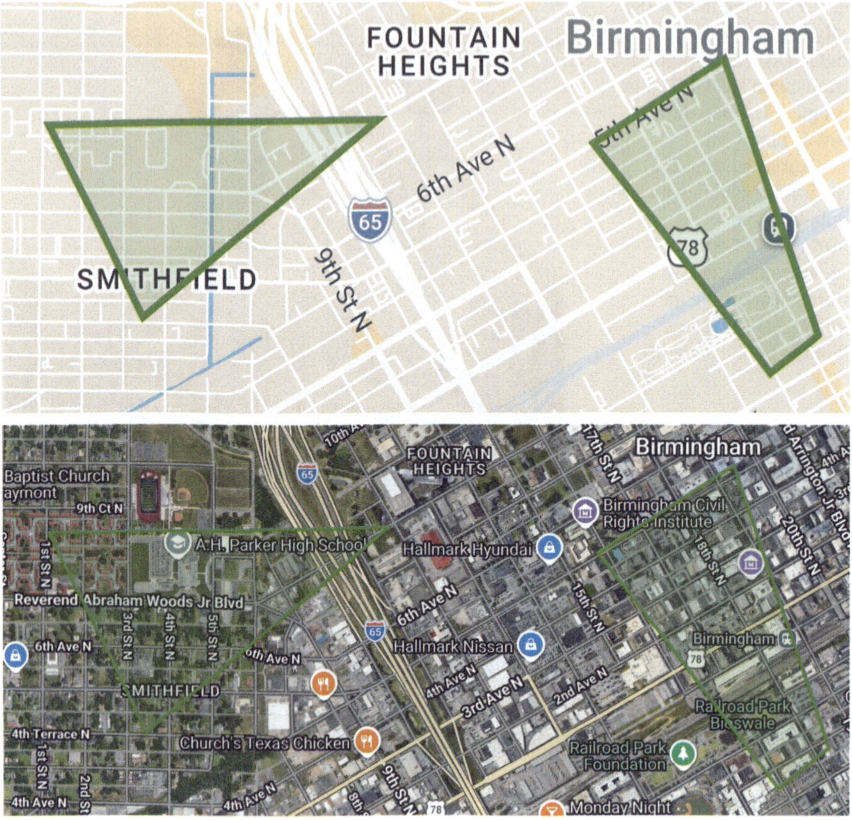

Fig. 15 Research-created map of Birmingham, indicating safe zones. Mrs. McDonald's Smithfield community (shape on the left) and the Fourth Historic District (shape on the right) (Google. (n.d.). Birmingham Safety Map. Retrieved April 16, 2025)

Integrating Neighborhoods

Black residents challenged the segregationist state laws that relegated them to inferior geographies and attempted to move to more desirable communities. In 1947 Black businessman Samuel Mathews purchased a lot in an area of Smithfield that was technically zoned as all-White. The developer who sold him the land felt confident that it would be imminently rezoned to Black, as it sat near an all-Black zoned section. However, after building his house, Mathews sought out and was denied an occupancy permit. Attorney Arthur Shores represented him in a successful lawsuit challenging the city's racial zoning. While Mathews won the lawsuit, he lost his home, which was bombed on August 18, 1947, the first of a wave of bombings in the neighborhood that were designed to strike terror in Black homebuyers and those who might sell to them (Eskew, 1997). From 1947 to 1963, more than 50 homes and buildings in predominately Black neighborhoods were bombed near the north end of Center Street in an area that came to be known as Dynamite Hill (Elliott, 2013). As a result of this domestic terrorism, Birmingham earned the nickname of "Bombingham."

The racial unrest and uncertainty shifted the city's demographics. Birmingham's 1960 population, according to Birmingham Public Library records, reached a census high of 340,887 residents, with 60.3% of them identifying as White. Then, the overall population began a continuous decline with White residents leaving the city as the Black population simultaneously increased (https://www.cobpl.org/resources/government/BirminghamPopulation.aspx, accessed March 1, 2025).

For a while, Ms. Cherry's family lived in Elyton Village Project, located near the imaginary racial dividing line on Center Street. Other Black families with greater means chose to integrate the mostly White Birmingham suburbs of Brookwood, Homewood, and Vestavia Hills. Ms. Cherry, a budding student artist at the time, safely explored more of her city after several court decisions deemed segregation unlawful. She told the story of visiting a friend who lived in a majority White neighborhood and the unsettling moment that followed:

> I was leaving his house, and that's when someone yelled, you know, 'N-word! What are you doing up here?' And it rattled me ... And I told him what happened, and he told his parents what happened ... But, you know, my visits up there after that were kind of limited.

According to records in the Birmingham Public Library (https://www.cobpl.org/resources/government/BirminghamPopulation.aspx, accessed March 1, 2025), the city's Black population steadily increased from 40% in 1950 to its peak of 74% in 2010. Richard Arrington became the city's first Black mayor in 1979. Although Birmingham has had its challenges, Mrs. McDonald is still proud to call it home.

> Black people not only need to hear the story, they need to learn and listen. They need to listen to the story, and they need to pay attention to what they had and what they went through to get it and learn how to appreciate it ... Birmingham is called the Magic City. And it is magic. It's magical because so many things have gone on in Birmingham, Alabama ... We're still here, and we're still doing things that are profitable to the Black community as well as other communities.

Edgecombe County, North Carolina: Terror, Exploitation, and Loss

Racial divisions in Edgecombe County, North Carolina, including in its county seat of Tarboro, were formed during plantation slavery. Rural Edgecombe was organized by farmland and property ownership, and the town of Tarboro, which used to be solely White, was organized as a center for related commerce (Mobley, 1986). Mrs. Lucy Whitehead, age 99, has lived in rural Edgecombe County her whole life. During her interview for this project, she shared a story that she learned from her grandmother, who "was in slavery." She described the ways enslaved children had to eat during the yearly festival:

Fig. 16 Princeville Mayor W. Ray Matthewson stands on the steps of the Princeville Community Center in the 1960s (Photo courtesy of Edgecombe Community Library; used with permission)

> [W]ith the grown people, they would have better food for them. But the children, they would have a trough. And she said, they put milk in there … and they put cornbread in there. And that's what the children had to eat. They had to put their mouths down in there and get their heads in there a little [and use a makeshift spoon].

Following the Civil War and abolition, many freed Black Americans stayed in Edgecombe County. Most remained on the farms that they had worked under bondage, now in the role of sharecropper or tenant farmer. However, some created the community of Princeville, North Carolina—established in 1885 just across the river from Tarboro; it is the oldest Black town in the United States (Mobley, 1986). At its inception, many White people in Tarboro and Edgecombe County favorably viewed Princeville as a way to keep Black citizens and workers nearby while not actually living within the town limits of Tarboro (Ibid.). For decades to follow, this enforced social distance remained a key feature in Tarboro and Edgecombe County. See Fig. 16 for an image of Princeville Mayor W. Ray Matthewson standing on the steps of the Princeville Community Center in the 1960s.

The Racial Divide

In mid-twentieth century Tarboro, public spaces and services were racially segregated, and separate water fountains were one manifestation of this divide. Mrs. Whitehead recalled a time when, as a young woman, she was confronted with the choice of using the "nasty" dipper designated for "colored" citizens or the other, "pretty and clean" dipper. She chose the latter, and a White man who had been watching approached her and her sister:

> [He said], "Didn't y'all see that dipper hanging there?"
> I said, "Yes."
> He said, "Didn't you see that sign that says colored?"
> I said, "Yes, I did, but it didn't say *what* color."
> That man got so mad at me. My sister said, come on, because your big mouth can get us in jail. But I told the truth.

Seventy-nine-year-old Mr. Matthew "Truman" Jenkins, a former farmer and Edgecombe community leader, recalled during his interview that some Black Edgecombe residents, including his grandfather, did not feel comfortable going to Tarboro, especially near dusk. Unlike most of Tarboro, however, the all-Black neighborhood of East Tarboro was understood to be a safe space for Black residents. Former East Tarboro resident Mrs. Stellyne Curtis, age 71, described life in her neighborhood as very positive, especially in the streets right around her house: "I was happy there."

Until the late 1960s, schools in Tarboro and Edgecombe County were legally segregated by race. Mrs. Whitehead remembered walking a long distance as a young person to get on a bus and ride to her all-Black school in Pinetops, a small town south of Tarboro. Figure 17 shows the 1944 faculty of all-Black Pattillo School in East Tarboro; Mrs. Whitehead worked at the school, and Mrs. Curtis fondly remembered her days there as a student.

Fig. 17 The 1944 faculty of all-Black Pattillo school in East Tarboro (Photo courtesy of Edgecombe County Memorial Library; used with permission)

Fig. 18 Combined rendering of safe and unsafe areas in Edgecombe County according to interviewees (Google (n.d.). Tarboro Safety Map)

Later, Mrs. Curtis was voluntarily part of Tarboro's school integration. Her experience at the formerly all-White Tarboro High was not positive:

> I went to Tarboro High, which I've always said was one of the biggest mistakes in my life. And I ask myself, why did I do it? And I know why I did it. I said I was going to be an ambassador, but all I was was hurt.

This racial animosity in Edgecombe County sometimes had deadly consequences, especially in Macclesfield, a small town south of Tarboro. As Mr. Jenkins put it: "they used to hang Black people down there." A cousin of his once saw a body hanging from a tree, and his uncle was forced to leave his home in Macclesfield to avoid a similar fate. This uncle had been accused of speaking inappropriately to a White woman, and family members found a way to help him safely leave town: "You know how they got him out? They piled corn on top of him … and shipped him to Newport News." Macclesfield had signs to intimidate Black people near the town, and, perhaps, warn them of this racist violence. One read "Run N***er Run." But Macclesfield wasn't alone in this white terror. Mr. Jenkins remembered seeing crosses burning in front of Black people's homes in rural Edgecombe.

Figure 18 is a combined rendering of the Edgecombe areas considered safe and unsafe by Mrs. Whitehead, Mr. Jenkins, and Mrs. Curtis. The orange areas were considered variably safe and welcoming. East Tarboro is the small green area within Tarboro. Pinetops, while a predominantly Black town, was considered unsafe by Mr. Jenkins. The small green box southwest of Tarboro was the area Mr. Jenkins felt most comfortable and safe—his own home. Unmarked areas were also considered variably safe and welcoming to Black citizens depending on who the citizen was, how well they were known, and how they responded to others. Macclesfield is marked in red at the bottom left corner of the map.

Entrepreneurship and Economic Factors

Edgecombe and its Black citizens have always relied on farming for economic growth: "most [Black] people … had their own little farm," said Mrs. Whitehead. After emancipation, sharecropping and tenant farming were common practices. Mr. Jenkins' father was a "sharecropper" on a White man's farm and turned "about 30 acres" until he passed away. After his father's death, Mr. Jenkins and his mother took over the sharecropping for a year before moving to his mother's father's land, where Mr. Jenkins still lives today. Figure 19 is an image circa 1940 of a mule-drawn baler in Edgecombe County being operated by Black farmers while White men look on.

While a minority, many Black people in Edgecombe County owned their own houses and land. Ownership, however, did not ensure equal access to farming supplies. Mrs. Whitehead and Mr. Jenkins both recalled one farm store, Clark Jenkins, with racist business practices. Mrs. Whitehead remembered that Black farmers had

Fig. 19 A mule-drawn baler in a field, Edgecombe County, N.C., circa 1940. Black farmers operate the baler while White men look on (Photo courtesy of Edgecombe County Cooperative Extension Office; used with permission)

to go to the back door when purchasing at the store. Mr. Jenkins told a story about a time his grandfather attempted to buy fertilizer but was denied because the store owner wanted to take over his grandfather's farm himself.

Eventually, much of Edgecombe's Black-owned farmland was purchased by White landowners. Mr. Jenkins explained the shift: "All the places where you're talking about now changed back in the '60s. It changed because people didn't make a lot, so they went north looking for jobs." As tractors and combines increased the speed of farming, land operators needed fewer workers and jobs disappeared. Furthermore, equipment was expensive, and Black farmers rarely had the capital to purchase the new machines.

The amount of Black land loss in Edgecombe County is staggering. In 1954, Black farmers fully owned 7580 acres in the county (U.S. Bureau of the Census, 1956). By 1997, only 341 acres were fully owned by Black farmers, representing a 95.5% total loss (U.S. Department of Agriculture, 1999). In Edgecombe, as across the United States, the Black experience of racial segregation and isolation has been shaped by the overlapping forces of spatial division and economic exploitation.

Conclusion

While each geographic place described in this chapter is entirely unique, Tulsa, Kinloch, Birmingham, and Edgecombe County all share a history marked by the systemic racial division of the mid-twentieth century United States. Black Americans in these areas lived in close proximity to yet completely separate from their White neighbors. The stories in this chapter share the particular experiences of 11 Black individuals who lived, worked, and moved within these parallel universes during and after de jure and de facto segregation. Their actions and geographies were variably constrained by their race, but, both inside and outside these boundaries, they built homes, churches, schools, and businesses—powerful communities marked by excellence and resilience.

When asked if they thought it important to preserve the history of geographic racial separation, almost all interviewees responded strongly in the affirmative. Their reasons varied; most wanted history, and more importantly, young people, to remember the ways in which these communities thrived and their residents took care of one another. Some also wanted to make clear the institutional racism that conditioned these communities' existence and challenges. To continue this work, the authors hope that other researchers employ similar phenomenological methods that engage individuals in telling their stories and mapping their geographic experiences.

Any such future endeavors might learn from some of this project's limitations. Finding interviewees who were old enough to vividly remember their pre-integration experiences and able to share those memories proved challenging, especially when trying to understand places with which the researchers themselves were not familiar. Many of the interviewees, in fact, explained that their parents—who are no longer living—should have been interviewed if one really hoped to understand the most important dynamics in these racially divided communities. Future researchers should set out quickly to conduct more research in this vein before these stories are lost to time.

Related projects might also seek out and incorporate more images of the signs and artifacts that enforced racial separation. For most of this project's interviewees, such signs were unnecessary and unknown; boundaries were community knowledge passed on from one generation to the next. Even for those signs that were known, such as the one in Macclesfield, NC, photographs proved impossible to find. Older individuals may recall more of these dividing markers, and researchers should consider incorporating questions regarding personal photographs or known public archives that may include such images. Similarly, photographic or geographic evidence of official historical racial boundaries could be gathered—for example, evidence of the school redistricting efforts and related geographic separation that played key roles during racial segregation in the United States.

Integration efforts and experiences might also provide powerful narratives for future work. In many of our interviews, participants described positive acts by White people that had an outsized impact on their young lives. And, while they were limited, some community spaces were designed to encourage integration and cross-racial connection. More stories about these moments and places could allow academics—and others—the opportunity to reconsider and refine anti-racist community practices.

Finally, future research might build upon this work by documenting through phenomenological methods contemporary racialized parallel universes in the United States. Younger people, perhaps even the grandchildren and great-grandchildren of the individuals interviewed here, have their own stories of separation to tell. While legal segregation has ended, its infrastructural impacts undoubtedly live on. And, only in knowing these stories, can we begin to build a different future.

Acknowledgements The authors extend their thanks to our interviewees: Stellyne Curtis, Dr. La Verne Ford Wimberly, Attorney James O. Goodwin, Matthew "Truman" Jenkins, Barbara McDonald, David Rice, Wilatrel Rice, James Simmons, Shirley Simmons, Lucy Whitehead, and LaDonna Wright Cherry. They also thank Brant Beene, Roslyn Croft, Pam Edmondson, Dr. Bob Ernst, Anne Gilliland, Dr. James H. Johnson, Jr., Reverend Richard Joyner, Alana Marie, Larissa McNeil, Gloria Morning, Mary Beth Newbill, Representative Raychel Proudie, Lauren Sallwasser, Jodie Hamer-Schlobohm, Clifton Taulbert, and Stephanie Willen Brown.

This research was approved by the University of North Carolina's Institutional Review Board.

References

Birmingham Times. (2020, February 6). A timeline of Birmingham's Black history. *The Birmingham Times*. https://www.birminghamtimes.com/2020/02/a-timeline-of-birminghams-black-history/

Brophy, A. L. (2003). *Reconstructing the dreamland: The Tulsa Riot of 1921: Race, reparations, and reconciliation*. Oxford University Press.

Catney, G., Frost, D., & Vaughn, L. (2019). Residents' perspectives on defining neighbourhood: Mental mapping as a tool for participatory neighbourhood research. *Qualitative Research, 19*(6), 735–752. https://doi.org/10.1177/1468794118803841

Coleman, A. K. (1999). *Socioeconomic analysis of the Greenwood District of Tulsa, Oklahoma: 1940-1980*. Oklahoma State University, Open Research Oklahoma. MA Thesis. https://openresearch.okstate.edu/server/api/core/bitstreams/a1d8bec1-c08a-4663-b21d-ac2c1ed45ee7/content

Elliott, D. (2013, July 6). Remembering Birmingham's Dynamite Hill neighborhood. Code switch [Audio Podcast]. *National Public Radio*. https://www.npr.org/sections/codeswitch/2013/07/06/197342590/remembering-birminghams-dynamite-hill-neighborhood

Eskew, G. (1997). *But for Birmingham*. The University of North Carolina Press.

Gillette, M. B., & Hurley, A. (2018). Vision, voice, and the community landscape: The Missouri Place Stories pilot project. *Landscape and Urban Planning, 173*, 1–8.

Google. (n.d.-a). Birmingham Safety Map. Retrieved April 16, 2025, from https://www.google.com/maps/d/u/0/edit?mid=1ERFNM1tLNBStQmL-Jz35aytQD5cf1Kk&usp=sharing

Google. (n.d.-b). Kinloch Maps—Municipal boundaries. Retrieved April 4, 2025, from https://www.google.com/maps/d/edit?mid=1VLkKlcWok865AszWEV8fxx7dCJ5B1NM&usp=sharing

Google. (n.d.-c). Kinloch Maps—Rice Safety Map. Retrieved April 4, 2025, from https://www.google.com/maps/d/edit?mid=1Uz_BDtog8C7d1VC8K1X06kWT-W4RErU&usp=sharing

Google. (n.d.-d). Kinloch Maps—Simmons Safety Map. Retrieved April 4, 2025, from https://www.google.com/maps/d/edit?mid=1UzHxxAGB5L0pUS4EMJRcTAMusjQQhpM&usp=sharing

Google. (n.d.-e). Tarboro Safety Map. Retrieved April 4, 2025, from https://www.google.com/maps/d/edit?mid=1kbcnwt-HGrrg2YDhgo96IPagyAvxQyM&usp=sharing

Google. (n.d.-f). Tulsa Safety Map. Retrieved April 16, 2024, from https://www.google.com/maps/d/edit?mid=1rc5cS9XCCSNlOBDddl_LBoL7DtM2ZVs&usp=sharing

Hanhorster, H. (2000). Whose neighbourhood is it? Ethnic diversity in urban spaces in Germany. *GeoJournal, 51*(4), 329–338.

Hemphill, E. (2020, May 6). Nearly complete "Kinloch doc" explores rich history of all-Black city "hanging on by a thread." *St. Louis on the Air. St. Louis Public Radio*. https://www.stlpr.org/show/st-louis-on-the-air/2020-05-06/nearly-complete-kinloch-doc-explores-rich-history-of-all-black-city-hanging-on-by-a-thread

Ike, N. (2024). Seeking justice in Black spaces: The geography, memory, and legacy of the Tulsa Race Massacre. *Historical Archaeology, 58*(2), 307–325.

Majors, T. (2022, November 12). Why is Birmingham called the magic city? *About Town*. https://www.abouttown.io/features/why-is-birmingham-called-the-magic-city

Marsh, B., Parnell, A. M., & Joyner, A. M. (2010). Institutionalization of racial inequality in local political geographies. *Urban Geography, 31*(5), 691–709.

Massey, D. (1991). A global sense of place. *Marxism Today*, 24–29. https://banmarchive.org.uk/marxism-today/june-1991/a-global-sense-of-place/

Messer, C., Beamon, K., & Bell, P. A. (2010). The Tulsa Riot of 1921: Collective violence and racial frames. *Western Journal of Black Studies, 37*(1), 50–60.

Missouri Census Data Center. (n.d.). *Missouri population 1900-1990*. University of Missouri Center for Health Policy. Retrieved March 24, 2025, from https://mcdc.missouri.edu/population-estimates/historical/cities1900-1990.pdf

Mobley, J. A. (1986). In the shadow of White society: Princeville, a Black town in North Carolina, 1886-1915. *North Carolina Historical Review, 63*(3), 340–384.

Moreno, C. (2021). Decades after the Tulsa Race Massacre, urban 'renewal' sparked Black Wall Street's second destruction. *Smithsonian Magazine*. https://www.smithsonianmag.com/history/black-wall-streets-second-destruction-180977871/

Pickles, J. (1985). *Phenomenology, science and geography: Spatiality and the human sciences*. Cambridge University Press.

Raanan, M. G., & Shoval, N. (2014). Mental maps compared to actual spatial behavior using GPS data: A new method for investigating segregation in cities. *Cities, 36*, 28–40.

Rey, S. J., & Knaap, E. (2024). The legacy of redlining: A spatial dynamics perspective. *International Regional Science Review, 47*(1), 3–44.

Rothstein, R. (2017). *The color of law: A forgotten history of how our government segregated America*. Liveright.

Schallhorn, K. (2020, June 24). Raychel Proudie is on a mission to clean up Kinloch—And she's just getting started. *The Missouri Times*. https://themissouritimes.com/raychel-proudie-is-on-a-mission-to-clean-up-kinloch-and-shes-just-getting-started/

Schneider, M. R. (2002). *"We return fighting": The civil rights movement in the Jazz Age*. Northeastern University Press.

Seamon, D. (2023). *Phenomenological perspectives on place, lifeworlds and lived emplacement: The selected writings of David Seamon*. Routledge.

Smiley, S. L. (2013). Mental maps, segregation, and everyday life in Dar es Salaam. *Tanzania. Journal of Cultural Geography, 30*(2), 215–244.

Stelzer, C. D. (2000, May 10). Promised land. *Riverfront Times*. https://www.riverfronttimes.com/news/promised-land-2474607

Sullivan, P. (2004). "To make America safe for democracy": Black freedom struggles in the World War I era. *Reviews in American History, 32*(1), 76–83.

Susaneck, A. P. [@segregation_by_design]. (2023a, July 17). The famous "Black Wall Street"—Before and after highway construction and "urban renewal" [image]. *Instagram*. https://www.instagram.com/p/CuzKdarM3X6/

Susaneck, A. P. [@segregation_by_design]. (2023b, August 15). Photos of Greenwood, Tulsa, before and after the 1921 Race Massacre [image]. *Instagram*. https://www.instagram.com/p/Cv92QE6MwnF/

Teague, H. A. (2019). Bullets and ballots: Destruction, resistance, and reaction in 1920s Texas and Oklahoma. *Great Plains Quarterly, 39*(2), 159–177.

The Oklahoma Eagle. (2025). About us. https://theoklahomaeagle.net/about-us/

Trounstine, J. (2018). *Segregation by design: Local politics and inequality in American cities*. Cambridge University Press.

U.S. Bureau of the Census. (1956). *United States census of agriculture: 1954. Vol 1, Part 16. North Carolina and South Carolina*. U.S. Department of Commerce. https://agcensus.library.cornell.edu/wp-content/uploads/1959-North_Carolina-1959-01-full.pdf

U.S. Department of Agriculture. (1999). *1997 census of agriculture: North Carolina state and county data. Volume 1, Part 33 (AC97-A-33)*. U.S. Department of Agriculture. https://agcensus.library.cornell.edu/wp-content/uploads/1997-North_Carolina-1997-01-intro-1.pdf

Weyeneth, R. R. (2005). The architecture of racial segregation: The challenges of preserving the problematical past. *The Public Historian, 27*(4), 11–44.

Wheeler, J. O. (1976). Transportation problems in Negro ghettos. In R. T. Ernst & L. Hugg (Eds.), *Black America: Geographic perspectives*. Anchor Books.

Wright, J. A., Sr. (2000). *Kinloch: Missouri's first black city*. Arcadia.

The Currents of Spiritual Capital: Expanding Normative Understandings of Water, Victimhood, and Agency in Freedmen's Towns

Cassandra R. Davis, Simona Goldin, and Iheoma Iruka

Introduction

Historical Black settlements, known as Freedmen's Towns, are on the verge of extinction in the United States. Newly freed Black people established Freedmen's Towns after the U.S. American Civil War as a sanctuary from racist and obstructive laws that severely constrained their life opportunities. To the town's founders, these communities represented spaces of possibility where Black people could commune, work, educate, and worship together safely. However, research has established that newly freed Blacks had limited access to hospitable land and were often forced to establish their towns in environmentally unstable spaces (Davis, 2023). The literature on the founding of Freedmen's Towns has uncovered how Blacks were often forced to settle on land that was undesirable to white people (Grace-McCaskey et al., 2021).

Scholar Andrea Roberts (2024) has documented over 500 Freedmen's Settlements in Texas. Others were identified in North Carolina (Davis, 2023), South Carolina (Zeise, 2012), Georgia (Botwick, 2018), and other states (Walton-Raji, 2023). Unfortunately, very few of these towns remain standing or exist today. Of those that remain, recent modeling reveals that Black communities overall have about a 26 percent increase in flood risk within the next 30 years due to climate change (Wing et al., 2022), leaving Freedmen's Towns highly susceptible to complete erasure.

C. R. Davis (✉) · S. Goldin
Department of Public Policy, The University of North Carolina, Chapel Hill, NC, USA
e-mail: cnrichar@email.unc.edu; simonag@email.unc.edu

I. Iruka
Gillings School of Global Public Health, The University of North Carolina, Chapel Hill, NC, USA
e-mail: iruka@unc.edu

S. D. Brunn (ed.), *Transdisciplinary Research Across Knowledge Worlds*,
https://doi.org/10.1007/978-3-032-19491-6_10

When the history of Freedmen's Towns establishments is told, it has largely been chronicled in ways that focus on newly freed Blacks' lack of agency in selecting their land (Cooper Jr., 2019; Davis, 2023; Grace-McCaskey et al., 2021; Mizelle Jr., 2016). These tellings highlight Black people's victimhood and lack of opportunity. For instance, Mizelle Jr. (2016, p. 34) asserted that a Freedmen's Town in North Carolina was constructed on land that was "discarded and unwanted" by white residents since it was prone to frequent and concurrent storms.

However, scholars have long noted the incompleteness or the danger of "single stories" (Adichie, 2009), revealing how singular issue stories are often incomplete. Thus, we ask: What if newly enslaved people had more agency and selected such places as an ode to their ancestors' past? Would that bring about more urgency in the renewal and revival of these historical places? Given that Freedmen Towns assumed foundational tragedy, are such places worth saving?

In this paper, we actively resist damage-centered research (Tuck, 2009) that focuses on historically minoritized people and centers deficits. Instead, we seek to present and understand systemic challenges as full and complete individual pathologies. We build upon a long line of scholarship that imagines parallel universes, that is, strength amidst challenge, resistance amidst racial injustice, and past amidst the present. We draw on the work of sociologist Sara Lawrence-Lightfoot (Lawrence-Lightfoot & Davis, 1997) who developed the research methodology of portraiture that was guided by the question: "What is good here?" and also Eve Tuck (2009), a scholar whose work theorizes damage-centered research and the possibilities of disruption. In addition, we draw on the work of Robert Hill (2003) whose seminal research on the strengths of Black families identified a strong religious orientation as one of five key strengths, each of which constitutes "adaptations necessary for survival and advancement in a hostile environment." In addition, our research builds on important recent scholarship that examines the strengths and capital of Black families (Clark et al., 2023) and charts a significant "rediscovery" of foundational texts that have seeded important strengths that are focused and culturally anchored research (Hill et al., 2023). These scholars and texts are in dialogue with critical race theorist Tara Yosso, whose community cultural wealth model (2005) is another conceptual framework that enables scholars to identify important resources and strengths in historically minoritized communities.

Together, these theoretical frameworks enable us to consider Black people's longstanding cultural and spiritual connections to water and complicate singular normative narratives of victimhood in the site selections of Freedmen's Towns. While Hill, Yosso, Lawrence-Lightfoot, and others' asset-based work has been widely used by sociologists studying schools and communities, these frameworks have not been leveraged in the geography literature. Doing so adds an important contribution to extant scholarship, leveraging theoretical and analytical tools to resist damage-centered, deficit-framed research and illuminates the concept of parallel universes.

In leveraging these theoretical frameworks, we argue that newly freed Black individuals drew on a form of spiritual capital when identifying and purchasing their land. Our analysis reveals parallel universes, shedding light on the ways in which Black people may have had more agency in acquiring their land than previously reported and white supremacy dictated how much of the land was available or could be purchased. We pose a powerful yet unasked question: To what extent did newly freed Black people choose, accept, and identify the sites for their towns? What informed their selection, and were there ways in which that selection was guided by their spiritual and cultural connections to water? Further, how, if at all, might past connections to Western African culture have informed Africans' selection of a new land? These questions invite us to rethink what a *just recovery* (Davis, Berke, et al., 2024) entails for these historical sites, regardless of the ongoing impacts of natural disasters. To arrive at these answers, we propose what Gloria Ladson-Billings (1998) refers to as *counterstories*, or naming one's truth by structuring our approach around the multiple ways water represents both death and life in historical artifacts and spoken word.

This paper contributes to the practice of cultural geographies by leveraging an important set of conceptual frameworks that, together, disrupt damage-centered analyses and enable more complete understandings of nuanced and rich histories. This research does not reduce the founders of Freedmen's Town and their generational families to a single story of victimhood. Instead, asset frameworks enable us to fully humanize these founders and families by foregrounding familial and cultural wealth in the face of systemic inequity, thereby understanding the effects of climate disasters and the possibilities of a just recovery. In the process, we contribute new understandings and counterstories that illuminate the interactions between cultural ways of being, geographic space, and climate disaster recovery. In the following section, we ground our work within the context of several asset-based frameworks.

A Grounding in Asset-Based Frameworks

What Is Good Here?

Sara Lawarence-Lightfoot (2019) delivered a keynote talk entitled "Forging the connections between learning and justice: On view, voice, and visibility" at the University of Michigan. In that presentation, Lawrence-Lightfoot invited educators, policymakers, and scholars to "challenge inequalities and asymmetries of education" by pivoting towards the light:

> Trying our hardest to figure out ways of describing and framing this challenge ... before us in ways that will not be negative and destructive, but rather will be imaginative, illuminating, enlightening.... Let's begin by asking the question, what is good here? What is resilient here? What is strong? What is worthy? And if we have a view that looks at what's good, not in a romanticized or idealized way, but rather in a way that allows us to see things that are strong and growing there. If we do that, we'll see a very, very different reality than what we've seen before.

Asset-based work can be dangerous—romanticized, pollyannaish—in which case equity work can become, in fact, weaponized (Goldin et al., 2021; Khasnabis & Goldin, 2020) if it is not built on critical appraisals of the endemic nature of racism in the U.S. (Anderson et al., 2022; Crenshaw, 1991; Delgado & Stefanic, 2001). Thus, building on the foundational assumptions of critical race theory, we proceed, as Hill did in 2003, with the fundamental knowledge that these key strengths are, in fact, "adaptations necessary for survival and advancement in a hostile environment." Our first layer in our asset-based framework is grounded in the ongoing quest to identify, highlight, and cherish what is good in situations that, on the surface, appear to be bad. A commitment to documenting strengths in the face of historic and contemporaneous injustice opens up a parallel universe by critically identifying that which may be difficult to identify.

Using asset-based frameworks enables us to examine and acknowledge parallel universes. Figure 1, a picturefrom Ironton, Louisiana, a Freedmen's Town and the oldest Black settlement in the state of Louisiana, assists with our conceptualization of wealth, asset-based framing, and capturing parallel universes. The photo shows oyster shells in a field, but it is also the contents of the sidewalk within the community. These shells could be a clear representation of underinvestment from the local governments since they were ground up and applied loosely as gravel throughout. However, residents recalled the beauty and strength of these shells and described how they connected homes and families, establishing a sense of community. One resident shared a remembrance of the sound these shells would make as children rushed between homes to visit friends. There, the presence of oyster shells represented a sense of connectivity and familiarity. Memories were evoked both visually

Fig. 1 Oyster shells ground up and applied loosely as gravel to make sidewalks in Ironton, Louisiana represent underinvestment from the local governments. (Photograph by Gary George)

and auditorily and with joy as residents focused on the historical connections within the community.

Community Cultural Wealth

To our knowledge, this account is the first time that critical asset-based frameworks have been leveraged to better understand the strengths, knowledge, and resources of communities of color in relation to climate disaster recovery. Bringing community cultural wealth into the analysis of a Freedmen's Town was particularly useful in its application to our study. To illustrate how community cultural wealth can be applied in the field of geography and disaster management, we selected quotes from former respondents of a Freedmen's Town in North Carolina (Davis, Dede, et al., 2024). The first quote is an example of aspirational capital—a way to hope and aspire to future goals. With repeated climate disasters battering, Princeville, North Carolina (the oldest incorporated Freedmen's Town in the U.S.), residents retained the ability to hold onto hope in the face of inequality and offer limited ways to make their dreams come true as they argue not for relocation but for rebuilding to promote tourism. A resident stated:

> Our goal is to become the world's number one tourist attraction. And we can do that just with our title ... to hopefully attract some people who can relocate here, build businesses, and also go to work, have fun, and play as well.

Similarly, we see social capital (i.e., systems of people and community resources that help individuals navigate through society) in a resident's recollection of how a family member opened their doors to assist with recovery during a flooding. A Freedmen's Town resident stated,

> We had another cousin working out of town and he said, 'You need a place to stay? My house is there, the lights are on, the water is on, you can stay there as long as you want.'

While often invisible, disregarded, or overlooked in disaster recovery, these exemplars powerfully recast the predominant deficit narrative about communities of color who experience climate disasters.

Figure 2 is a picture of a journal created by a resident of a Freedmen's Town in Louisiana. These pages are filled with the testimony and witness perspective of life in the community after Hurricane Ida. While the storm destroyed most homes, this community member documents the history and present of the community through writings. The words highlight the strength and great obstacles the town will face in the future for recovery and stand as a powerful exemplar of the "supportive network and optimism for challenges" factor, validated by Iruka et al. (2025). This image could be seen as someone journaling about their hardships in the aftermath of a traumatic event. And yet, this photo reflects the power of individuals as memory workers documenting current events to address one's ability to survive and thrive post-disaster.

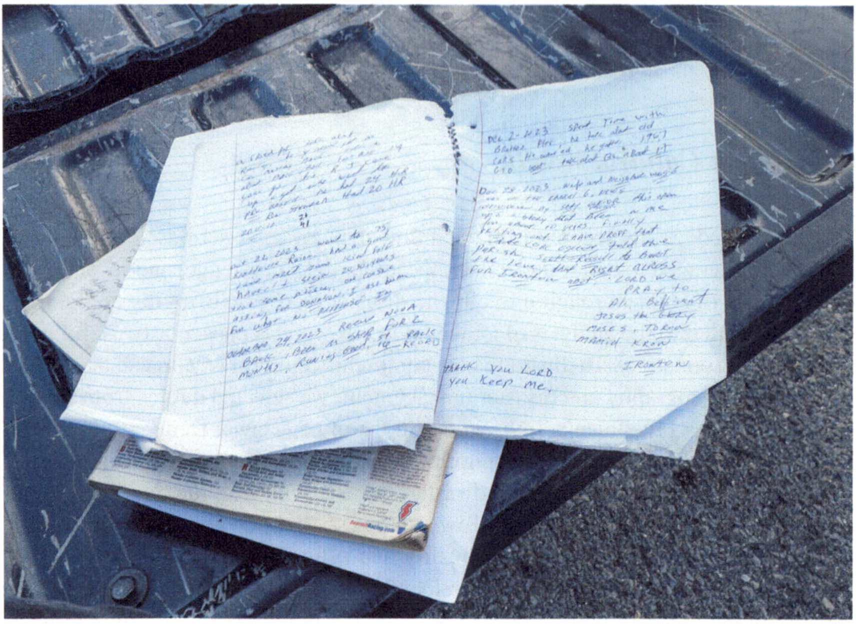

Fig. 2 Pages from an individual's journal after Hurricane Ida, reflect the power of individuals as memory workers documenting current events and the ability to survive and thrive post-disaster. (Photograph by Gary George)

In this example of resistant capital, we hear the spiritual significance of the land site as the speaker names the "sacred land" that Princeville sits on and how inextricably that land is tied to the community's identity. These examples highlight another form of strength, spiritual capital. In the next section, we illuminate the spiritual capital (e.g., Clark et al., 2023; Hill, 2003; Iruka et al., 2025; Neal-Stanley et al., 2024).

Spiritual Capital

Recent scholarship documents that spiritual capital encompasses the understanding of how spirituality and religion are another form of asset-based framing, especially for people of color (Clark et al., 2023; Hill, 2003; Iruka et al., 2025; Park et al., 2020). Robert Hill's work on the strengths of Black families has been studied by scholars to elaborate the importance of spiritual capital in Black families, while also highlighting the navigational skills and practices that Black youth bring to their work. Hill also identified key strengths characteristic of Black families, such as strong kinship bonds, adaptability in family roles, and a strong achievement orientation.

Pérez Huber (2009) argued that spiritual capital refers to a dependence on a higher power (i.e., a deity), as well as a retelling of ancestral practices and beliefs passed down through generations. In essence, this method powerfully connects individuals to the traditions of their community, family, and themselves. Through spiritual capital, connectivity is vital and can be achieved by building and maintaining relationships particularly within the Black church (Park et al., 2020).

Historical examples revealed the church as the bedrock of Black communities and established it as the central point of social change (Billingsley, 1999). Black churches were also deemed safe havens throughout the Jim Crow Era, a place where Black people could commune through spiritual and physical protection from white supremacy (Gates & Hoffman, 2021). There were even places that pushed against displacement, the gentrification of Black communities, community-based redevelopment, and civil rights (Born et al., 2021).

We also see examples of Black churches promoting mental well-being and health within their communities, particularly those residing in disinvested areas (Richardson et al., 2024). Black churches have provided congregants with essential support when dealing with significant losses (Bailey et al., 2024). Additionally, research provides examples of Black churches establishing pop-up vaccination sites for the flu (Nuss et al., 2024) and COVID-19 (Bateman et al., 2024) as well as serving as a source for distributing information about HIV (Husbands et al., 2021).

Given their symbolic, protective, and healing nature, these spiritual institutions have consistently been targets of oppressive attacks (Gates & Hoffman, 2021; Houdek, 2018; McCarthy, 2006). Gates and Hoffman (2021) accounted for the ways in which the Ku Klux Klan focused on dismantling and destroying Black churches during the Jim Crow era. Unfortunately, recent studies have shown that these attacks persist into the twenty-first century. A study conducted by Timothy McCarthy (2006) illustrated that during the last decade of the 20th century, roughly 500 Black churches were burned and mostly in the South.

Two images of Saint Paul Missionary Baptist Church in Ironton, LA, further illustrate spiritual capital (Figs. 3 and 4). In interviews conducted by our research team in 2024, Ironton's residents recalled memories of their relatives singing solos at the front of the sanctuary, making community-wide announcements that brought news of upcoming events, and highlighting celebrations of love and service (Davis, Berke, et al., 2024). For decades, generations of Ironton's residents attended the church, planted by their ancestors in the middle of their community.

Figure 4 is the interior picture of the church after Hurricane Ida's landfall in 2021. The storm wreaked havoc in the community, drastically reducing the town from 51 homes to 10 and forcing tens of families to leave their history and homes behind. The photo was taken in January 2024 with remnants still visible.

Through our critical asset-based framework, we see the power of the question: "What is good here?" (Lightfoot, 1986). Communities long thought of as the most vulnerable, marginalized, and under-resourced have strengths, resources, and community cultural wealth. Systemic inequality obscures assets and strengths to the detriment of disaster recovery. Our framework serves as a tool to challenge deficit views of communities and move towards a *just recovery*. It is one that not *only* sees

Fig. 3 Saint Paul Missionary Baptist Church in Ironton, Louisiana. (Photograph by Gary George)

Fig. 4 Inside Saint Paul. (Photograph by Gary George)

a destroyed institution (Fig. 3) that needs to be dismantled. Rather it is a framework that enables observers to see the vibrancy and resilience of generations of Ironton's residents who maintain permanence and connection even beyond disastrous events.

The spiritual capital of the Black community is undeniable. Although spirituality can be connected to religious institutions, individuals can derive spiritual benefits without exhibiting traditional religious behavior (McNamara Barry et al., 2010). In essence, individuals have their own autonomy in deciding their spiritual capital (Park et al., 2020).

In the following sections, we utilize our critical asset-based framework to provide a plausible explanation on the identification, selection, and purchase of Freedmen's Towns near water sources. We address varying representations of water to help provide a larger context around its connection to the Black community and Freedmen's Towns.

The Standard Universe: Following the Normative Discourse on Water

The most precise example of water equating death begins with the Trans-Atlantic slave trade where over 12.5 million Africans were kidnapped from their land and forced into bondage that would last over 400 years (The Trans-Atlantic Slave Trade Database, 2021). Captive Africans were forcibly transported in tight quarters and given minimal sustenance holding them for the journey across the Atlantic Ocean (Mustakeem, 2016). The seas were murderous. Africans journeyed across the waters in the harshest and most brutal of ways of individual and generational enslavement. Some survived the brutal journey. For The slave merchants used the oceans as a dumping ground to murder and dispose of bodies of enslaved Africans who were deemed deviant or diseased (Piersen, 1977).

Enslavers used the vast depth and width of the Atlantic Ocean to control their captives. Pierson (1977) provides a documented account of shipmates threatening enslaved Africans with disembodiment and throwing their limbs into the sea where they would be lost for eternity. The enslavers were aware that the cultural and spiritual beliefs of some they had kidnapped believed that one's whole body needed to be intact to rest in the afterlife. Thus, the enslavers leveraged this knowledge as another way to torment and dehumanize their captives by showing that they controlled their captives' bodies, lives, deaths, and beyond.

Water remained an essential tool for the murderous pursuit of enslavement during chattel slavery in the United States with continued use of water, such as rivers and harbors, to transport the people who were enslaved to communities situated inland (Parry, 2018) and to transport the very goods collected and produced by the enslaved (Johnson, 2013). Tyler Parry (2018) wrote that the Mississippi River was key in maintaining and expanding African enslavement by delivering the enslaved to ports for purchase. Walter Johnson (2013) argued that various waterways were also used to distribute products produced by chattel slavery, like cotton, rice, and sugar, worldwide which led to a thriving capitalist economy for planters and merchants.

In this way, enslavers continued to use water to fill their coffers and to expand their economic opportunities. The ubiquity and continuing horror of the term "sold down the river" encapsulates the power and threat of water, viz., to break families and communities of the enslaved, to punish, and to maximize enslavers' wealth. To this day, "sold down the river" refers to an ultimate and profound betrayal.

Following the end of slavery newly freed Africans began building churches, schools, and communal infrastructure which they sought to do by pursuing the purchase of land. As they did so, they sought to flourish—to pray, to learn, to support and to grow—and to be safe from White supremacy. The Freedmen's Towns were often situated in areas prone to disasters, such as flooding, hurricanes, and other disasters. For instance, Princeville's topography revealed a massive swamp directly adjacent to the Tar River (Cooper Jr., 2019). Furthermore, records show that the Tar River flooded at least six times over the past 100 years (1865, 1889, 1919, 1924, 1940, and 1958). Black residents continuously lost their family members, animals, homes, schools, and personal effects to the flooding waters with little support to encourage recovery (Davis, Berke, et al., 2024).

Water—in all its destructive and life-giving ways—continues to be the focus of many Black artists, storytellers, and musicians, as Black artists have written and sung about the numerous tragedies surrounding water (Granger, 2021). Bessie Smith (1927) from *Back-Water Blues* recounts the flooding of the Cumberland River, "When it thunders and lightin' and the wind begins to blow…there's thousands of people … aint' got no place to go." In the poem Aunt Su's Stories, Langston Hughes (2001, p. 40) wrote,

> Aunt Sue cuddles a brown-face child to her bosom
> and tells him stories.
> Black slaves
> Working in the hot sun,
> And black slaves
> Walking in the dewy night,
> And black slaves
> Singing sorrow songs on the banks of a mighty river…

In these and other tellings, we see water as harm, conductor, and enabler of Black pain. However, the story is broader than this, as water, in these counterstories, is also a purifying and renewing force. There is a rich history that dates to a time when water was equated with life and sustainability. We will explore these ideas next.

Introducing a Parallel Universe and Leveraging Counterstories

Before the Atlantic slave trade, West African communities were typically located near water, which led to their expertise in swimming, canoeing, diving, and capturing food (Dawson, 2018). The distance between Angola and Senegal spans thousands of miles of coastline with additional thousands of miles of rivers and other waterways. Thus, the customs, faiths, and social interactions of Western African communities were deeply intertwined with their lived experiences on land and in water (Dawson, 2018, 2023).

Water also occupied an important space of spiritual connection for West Africans, bridging between the past and present (Dawson, 2023; Piersen, 1977; Talkeu Tounouga & Brock, 2003). Water was sacred and integral to culture (Dawson, 2023; Talkeu Tounouga & Brock, 2003). Talkeu Tounouga and Brock (2003) have argued that water within the Sub-Saharan regions was seen as a source of life, renewal, restoration, and cleansing. Even these statements are the construction of a powerful *counterstory*, remaking singular views of the continent that do little to name the rich cultural traditions and expert knowledges of those who were kidnapped for enslavement.

The *Kalunga* is a concept originating from people in the Congo area and is believed to be the liminal space between two worlds, referring to a line within the Atlantic Ocean that permits the living and the dead to cross over (Parry, 2018). This idea was carried across the oceans by enslaved Africans to the United States. Matriarchal deities within the water, known as Mami Wata (Drewal et al., 2008) and Yemanja (Shirey, 2009), played pivotal roles as cultural guardians and water goddesses. These guardians and goddesses continue to be honored in modern-day celebrations held outside West Africa, specifically in Brazil (Shirey, 2009).

Enslavers recollected many instances of captured Africans jumping off ships, freeing themselves from bondage (Philips 1732). The enslaved believed water held their deities and ancestors and could help them return home (Dawson, 2023; Piersen, 1977). Furthermore, even after arriving in the New World, they believed that smaller bodies of water could connect them to the Atlantic Ocean, thus allowing them to return to their homeland. Pierson (1977, p. 153) quotes an enslaved woman in Georgia, U.S., who asserted that she would use the river as a means of escape, "[I'm going to] march right down in the river to take back to Africa." Years later, Langston Hughes (2001, p. 36) wrote *The Negro Speaks of Rivers* in which the speaker recalls their longstanding connection to rivers. The piece starts, "I've known rivers: I've known rivers ancient as the world and older than the flow of human blood in human veins." Where water was used as a symbol of death, it also held deep, resonant spiritual meaning as a symbol of life.

We write here about the power of counterstories and their contributions to narrative justice. Counterstories enable us to hear from and center the voices of those who have been silenced, thereby gaining a deeper understanding of a multiplicity of experiences. In this way, even hazards, such as flooding, which cause significant damage and loss, can also revive the land, providing a necessary baptism. These understandings are historical as research has shown that flood-based farming systems are prevalent in Sub-Saharan Africa where floods are a steady occurrence, not harmful, and create the foundation for a beneficial farming system (Puertas et al., 2021). These farming practices impact livestock, produce, and fishing. We do not claim that climate-induced disasters, particularly those centered on flooding, benefit Africa. The point is that enslaved Africans' earlier exposure and dependence on flooding for agriculture before being taken across the Atlantic would suggest a plausible interest in choosing a Freedmen Town that is situated in a space with flooding capabilities.

As such, we do not refute the violence of enslavement. Instead, we interrogate these histories through critical, asset-based lenses and construct powerful counternarratives that complicate and broaden our understanding of Freedmen's Towns as spaces of important, influential, and historic community cultural wealth. Additionally, our framing allows for Freedmen's Towns to have more agency and ingenuity. These seemingly contradictory meanings coexist, forming parallel universes that must be known, honored, and acknowledged.

Discussion

A fundamental argument of Yosso's community cultural wealth model is that communities and individuals exist in a society that features endemic racism, and that endemic racism makes invisible community cultural wealth. Thus, using community cultural wealth and other scholarship on the strengths that communities that have been historically and contemporarily marginalized, scholars have sought to push against majoritarian understandings of people of color's navigation through systemic racism and institutions meant to obstruct their access as simple victimhood. Instead, asset-based, critically informed frames allow social scientists to observe the resistance alongside the inequities and injustices of racial injustice. For instance, resistant capital comprises skills related to resisting oppression that have often been seen, in majoritarian views, as acts of non-compliance worthy of—and often receiving—punishment. Our analysis falls along these lines, recasting and reframing the physical siting of Freedmen's Towns. We see the ways that the evaluation of their siting is fully linked to white supremacy, viz., that Freedmen's Town founders had no choice and were compelled to establish their towns in areas that they did not want, a story that is incomplete. Resistance to white supremacy, cultural knowledge, and beliefs informed the management, survival, and thriving of communities even as they were impacted by white supremacy.

Water evoked both death and life to newly freed Black people in the US. Given these nuanced, historical, spiritual, and multifaceted understandings, we also suggest a counterstory re-telling or reconsideration of formerly enslaved peoples' site selections of Freedmen's Towns near water: Considering and advancing the possibility that those who established their towns, fortified by aspirational, navigational, and spiritual capital (amongst other forms of capital), may have also been purposeful. Seen through critical asset-based frames, we can observe how these choices were also agentic—sought after, meaningful, spiritually rich—while also the result of white supremacy and its brute force and violence. To those freed, water offered the possibility of returning home, a place where their ancestors recalled and reminded their descendants not to forget. Water held their former family members who danced with spirits like Mami Wata, Yemanja, and others. In addition, water could provide great fortune and allow them to prosper in a new land. These parallel universes exist alongside each other obscured by white supremacy and named and documented with critical asset-based frames.

While understanding that these places were likely selected due to deep meaning for life prior to enslavement, how much more important is it to ensure these places are not destroyed? We noted earlier that many have asked the question: Are Freedmen's Towns worth saving? We join the residents of Freedmen's Town in answering, unequivocally yes. We turn to the words of a Princeville resident:

> If you move the town of Princeville, then it's not the town of Princeville anymore. These are the sacred grounds where our forefathers shed their blood, sweat, and tears. And this is what we stand firm on.

However, we understand that *saving* can take many different forms. The work presented here builds on our previous work delineating principles of a *just recovery* (Davis, Berke, et al., 2024), one that acknowledges previous harm, celebrates the interconnectedness of the past and future, creates spaces for descendants of Freedmen's Towns to excel and justly recover. It is also informed by the voices and knowledge of Freedmen's Town residents. Any just recovery would honor and seek to build upon the community's cultural wealth. Efforts should be made to ensure that these places are restored and honored while preserving the history, present, and future of those who worked to build a new home.

References

Adichie, C. N. (2009, July). The danger of a single story [Video]. *TedGlobal*. https://www.youtube.com/watch?v=D9Ihs241zeg

Anderson, R. E., Heard-Garris, N., & DeLapp, R. C. T. (2022). Future directions for vaccinating children against the American endemic: Treating racism as a virus. *Journal of Clinical Child and Adolescent Psychology: The official journal for the Society of Clinical Child and Adolescent Psychology, American Psychological Association, Division 53, 51*(1), 127–142. https://doi.org/10.1080/15374416.2021.1969940

Bailey, A., Boakye, P., Esak, N., Johar, M., Velasco, D., & Leung, C. (2024). Driven by purpose: Exploring spirituality in black mothers' coping with loss of children to gun homicide. *Homicide Studies, 29*(1), 45–65. https://doi.org/10.1177/10887679241285908

Bateman, L. B., Hall, A., Hannon, L., Ryan, M., Osborne, T., Whitfield, S., Okoro, G., Stager, C., Driggers, S., Jones, V., Rhinehart, J., & Fouad, M. N. (2024). Partnering with churches to address COVID-19 vaccine hesitancy and uptake in trustworthy contexts. *American Journal of Public Health, 114*(Suppl. 5), S392–S395. https://doi.org/10.2105/AJPH.2024.307683

Billingsley, A. (1999). *Mighty like a river: The black church and social reform.* Oxford University Press.

Born, B., Berney, R., Baker, O., Jones, M. R., King, D., & Marcus, D. (2021). Pushing back on displacement: Community-based redevelopment through historically black churches. *Societies, 11*(1), 10. https://doi.org/10.3390/soc11010010

Botwick, B. (2018). Gullah-Geechee settlement patterns from slavery to freedom: Investigation of a Georgia plantation slave quarter. *North American Archaeologist, 39*(3), 198–228. https://doi.org/10.1177/0197693118793795

Clark, E. M., Ma, L., Williams, B. R., Ghosh, D., Park, C. L., Schulz, E., Woodard, N., & Knott, C. L. (2023). A longitudinal study of social, religious, and spiritual capital and physical and emotional functioning in a national sample of African-Americans. *Journal of Community Psychology, 51*(3), 978–997. https://doi.org/10.1002/jcop.22936

Cooper, J. T., Jr. (2019). Reflections on engaging socially vulnerable populations in disaster planning. In M. K. Lindell (Ed.), *The Routledge handbook of urban disaster resilience: Integrating mitigation, preparedness and recovery planning* (pp. 395–407). Routledge. https://doi.org/10.4324/9781315714462-24

Crenshaw, K. (1991). Race, gender, and sexual harassment. *Southern California Law Review, 65*, 1467.

Davis, C. R. (2023). The destructive long-term impact of natural disasters on Black and Brown schooling communities in the USA. In G. Noblit (Ed.), *Oxford research encyclopedia of education* (pp. 1–22). Oxford University Press. https://doi.org/10.1093/acrefore/9780190264093.013.1883

Davis, C. R., Berke, P., Hino, M., Goldin, S., Fetaw, R., Cornette, S., Lacey, M., & Morgan, T. (2024). *A landscape study of social equity data needs and its access and availability to support the disaster resilience of marginalized communities.* Coastal Resilience Center. https://naturalhazardsresiliencecenter.org/wp-content/uploads/2024/09/CRC-Landscape-Year-2-Report-Final.pdf

Davis, C. R., Dede, B., Morgan, J., Fetaw, R., George, G., Lacey, M., & Morgan, T. (2024). *A just recovery, weathering the storm in a Freedmen's town: An exploration of residents' cultural resilience through defiance.* https://www.coastalresiliencecenter.org/social-equity

Dawson, K. (2018). *Undercurrents of Power: Aquatic culture in the African diaspora.* University of Pennsylvania Press. https://doi.org/10.9783/9780812294781

Dawson, K. (2023). Waterscapes and wet bodies: Beach culture in Atlantic Africa and the diaspora, 1444–1888. *Coastal Studies & Society, 2*(1), 58–81. https://doi.org/10.1177/26349817231165157

Delgado, R., & Stefanic, J. (2001). *Critical race theory: An introduction.* New York University Press.

Drewal, J. H., Houlberg, M., Jewsiewicki, B., Nunley, J. W., & Salmons, J. (2008). *Mami Wata: Arts for water spirits in Africa and its diasporas.* Fowler Museum at UCLA.

Gates, H. L., & Hoffman, D. (2021). *The black church: This is our story, this is our song. Audio.* Penguin.

Goldin, S., Duane, A., & Khasnabis, D. (2021). Interrupting the Weaponization of Trauma-Informed Practice: "… Who Were You Really Doing the 'Saving' for?". *The Educational Forum, 86*(1), 5–25. https://doi.org/10.1080/00131725.2022.1997308

Grace-McCaskey, C. A., Pearce, S. C., Harris, L., Corra, M., & Evans, K. J. (2021). Finding voices in the floods of Freedom Hill: Innovating solutions in Princeville, North Carolina. *Journal of Environmental Studies and Sciences, 11*(3), 341–351. https://doi.org/10.1007/s13412-021-00701-5

Granger, C. (2021). Speaking of rivers: Spirituals, poetry, and relationality. *The Langston Hughes Review, 27*(2), 188–214. https://doi.org/10.5325/langhughrevi.27.2.0188

Hill, N. E., Pinderhughes, E. E., Hughes, D. L., Johnson, D., Murry, V. M., & Smith, E. P. (2023). African American family life: Rediscovering our foundational works. *Research in Human Development, 20*(3–4), 158–182. https://doi.org/10.1080/15427609.2024.2310453

Hill, R. B. (2003). *The strengths of Black families*. Bloomsbury Publishing PLC.

Houdek, M. (2018). Racial sedimentation and the common sense of racialized violence: The case of black church burnings. *Quarterly Journal of Speech, 104*(3), 279–306. https://doi.org/10.1080/00335630.2018.1486035

Huber, L. P. (2009). Challenging racist nativist framing: Acknowledging the community cultural wealth of undocumented Chicana college students to reframe the immigration debate. *Harvard Educational Review, 79*(4), 704–730. https://doi.org/10.17763/haer.79.4.r7j1xn011965w186

Hughes, L. (2001). *The collected works of Langston Hughes*. University of Missouri Press.

Husbands, W., Kerr, J., Calzavara, L., Tharao, W., Greenspan, N., Muchenje-Marisa, M., Luyombya, H., Nakamwa, J., Arnold, K., Nakiweewa, S., & Browne, O. (2021). Black PRAISE: Engaging Black congregations to strengthen critical awareness of HIV affecting Black Canadian communities. *Health Promotion International, 36*(2), 303–312. https://doi.org/10.1093/heapro/daaa057

Iruka, I. U., Forte, A. B., Liu, S., Sims, J., & Curenton, S. M. (2025). Initial validation of the family cultural wealth survey: Relation with racial discrimination and well-being for Black families. *Adversity and Resilience Science, 6*(1), 1–17. https://doi.org/10.1007/s42844-024-00139-y

Johnson, W. (2013). *River of Dark dreams: Slavery and empire in the cotton kingdom*. Harvard University Press. https://doi.org/10.2307/j.ctvjsf5q7

Khasnabis, D., & Goldin, S. (2020). Don't be fooled, trauma is a systemic problem: Trauma as a case of weaponized educational innovation. *Bank Street Occasional Paper Series, 43*, 5.

Ladson-Billings, G. (1998). Just what is critical race theory and what's it doing in a nice field like education? *International Journal of Qualitative Studies in Education, 11*(1), 7–24. https://doi.org/10.1080/095183998236863

Lawrence-Lightfoot, S. (2019, November 18). *Forging the connections between learning and justice: On view, voice, and visibility* [TeachingWorks streaming seminar]. TeachingWorks Seminar Series. University of Michigan School of Education.

Lawrence-Lightfoot, S., & Davis, J. (1997). *The art and science of portraiture*. Jossey-Bass.

Lightfoot, S. L. (1986). On goodness in schools: Themes of empowerment. *Peabody Journal of Education, 63*(3), 9–28. https://doi.org/10.1080/01619568609538522

McCarthy, T. P. (2006). A test of faith black church burnings and America's enduring crucible of "race." *Souls, 8*(1), 12–26. https://doi.org/10.1080/10999940500517023

McNamara Barry, C. M., Nelson, L., Davarya, S., & Urry, S. (2010). Religiosity and spirituality during the transition to adulthood. *International Journal of Behavioral Development, 34*(4), 311–324. https://doi.org/10.1177/0165025409350964

Mizelle, R. M., Jr. (2016). Princeville and the environmental landscape of race. *Open Rivers: Rethinking Water, Place and Community, 4*, 34. https://openrivers.lib.umn.edu/article/princeville-and-the-environmental-landscape-of-race-2/

Mustakeem, S. M. (2016). *Slavery at sea: Terror, sex, and sickness in the middle passage*. University of Illinois Press. https://doi.org/10.5406/illinois/9780252040559.001.0001

Neal-Stanley, A. M., Morgan, J. C., & Allen, D. J. (2024). The religio-spiritual capital of the Black Church: A conceptual model for combatting antiblackness in the early years. *Early Childhood Research Quarterly, 69*, S118–S128. https://doi.org/10.1016/j.ecresq.2024.03.003

Nuss, H., Privor-Dumm, L., Ukachukwu, C., & Hall, L. L. (2024). Building and sustaining flu vaccine acceptance and trust in the black community through partnerships with churches, salons, and barbershops. *Journal of Racial and Ethnic Health Disparities*. [Advance online publication]. https://doi.org/10.1007/s40615-024-02141-7

Park, J. J., Dizon, J. P. M., & Malcolm, M. (2020). Spiritual capital in communities of color: Religion and spirituality as sources of community cultural wealth. *The Urban Review, 52*(1), 127–150. https://doi.org/10.1007/s11256-019-00515-4

Parry, T. (2018). The role of water in African American history. *Black Perspectives*. https://www.aaihs.org/the-role-of-water-in-african-american-history/#:~:text=It%20was%20the%20channel%20through,to%20toil%20under%20King%20Cotton

Piersen, W. D. (1977). White Cannibals, black martyrs: Fear, depression, and religious faith as causes of suicide among new slaves. *The Journal of Negro History, 62*(2), 147–151. https://doi.org/10.2307/2717175

Puertas, D. G., van Steenbergen, F., Haile, A. M., Kool, M., & Embaye, T.-A. G. (2021). *Flood based farming systems in Africa*. Spate Irrigation Network Foundation. https://floodbased.org/wp-content/uploads/2021/05/OP_05_Flood-based-farming-in-Africa_SF.pdf

Richardson, B. T., Jackson, J., Marable, G., Barker, J., Gardiner, H., Igarabuza, L., Leasy, M., Matthews, E., & Zisman-Ilani, Y. (2024). The role of black churches in promoting mental health for communities of socioeconomically disadvantaged black Americans. *Psychiatric Services, 75*(8), 740–747. https://doi.org/10.1176/appi.ps.20230263

Roberts, A. (2024). *"The Texas Freedmen Colonies Project"*. https://www.thetexasfreedmencoloniesproject.com/

Shirey, H. (2009). Transforming the orixás: Candomblé in sacred and secular spaces in Salvador Da Bahia, Brazil. *African Arts, 42*(4), 62–79. https://doi.org/10.1162/afar.2009.42.4.62

Smith, B. (1927). *Back-water blues [Song]*. Columbia Records.

Talkeu Tounouga, C. T., & Brock, O. (2003). The symbolic function of water in sub-Saharan Africa: A cultural approach. *Leonardo, 36*(4), 283. https://doi.org/10.1162/002409403322258664

The Trans-Atlantic Slave Trade Database [estimates]. (2021). *Slave Voyages*. http://www.slavevoyages.org/tast/assessment/estimates.faces

Tuck, E. (2009). Suspending damage: A letter to communities. *Harvard Educational Review, 79*(3), 409–428.

Walton-Raji, A. (2023). Freedmen settlements of Indian Territory and three freedmen community clusters. *Great Plains Quarterly, 43*(3), 253–265. https://doi.org/10.1353/gpq.2023.a918406

Wing, O. E. J., Lehman, W., Bates, P. D., Sampson, C. C., Quinn, N., Smith, A. M., Neal, J. C., Porter, J. R., & Kousky, C. (2022). Inequitable patterns of US flood risk in the Anthropocene. *Nature Climate Change, 12*(2), 156–162. https://doi.org/10.1038/s41558-021-01265-6

Yosso, T. J. (2005). Whose culture has capital? A critical race theory discussion of community cultural wealth. *Race Ethnicity and Education, 8*(1), 69–91. https://doi.org/10.1080/1361332052000341006

Zeise, J. L. (2012). *Dawn of freedmen: The Freedmen's town of Mitchellville on Hilton Head Island, S.C.* University of South Carolina, History Department. Unpublished PhD dissertation.

The Norwegian Paradox: Exploring Climate Leadership, Indigenous Sámi Voices, and the Energy Economy

Karen S. Barton

Introduction

The interplay between environmental progressiveness and economic dependence on fossil fuels presents a challenging set of contradictions for nations as they pursue sustainability agendas. Norway epitomizes this paradox, as it is celebrated globally for its forward-thinking climate leadership while remaining one of the largest exporters of oil and gas in the world. This "Norwegian paradox" raises fundamental questions about how countries reconcile economic growth with environmental stewardship. Using Julian Agyeman's concept of "just sustainability," which emphasizes the integration of social justice into sustainability (Agyeman, 2013), this chapter examines Norway's energy policies with a special focus on their implications for the Indigenous Sámi people. Drawing on 6 weeks of fieldwork conducted in 2022 as part of my Fulbright Hays fellowship to Norway, as well as a three-week follow-up visit in January 2024 to Sámi lands, this work investigates Norway's dual identity as a climate leader and fossil fuel exporter (Barton, 2025).

By also exploring the impacts of wind farm development on Sámi lands, the governance of Norway's sovereign wealth fund, and the nation's broader energy strategies, this chapter outlines the ethical challenges involved in balancing a green transition with the rights of Indigenous communities. Through specific case studies—including the Fosen Vind project and the Alta Dam—this chapter reveals the parallel universes that exist between those who live in mainstream Norway and those who dwell in the Arctic (Sara et al., 2021). These examples illustrate how vastly different priorities and lived experiences coexist and emphasize the critical need to prioritize equity and justice in energy policy so we can better bridge these divides.

K. S. Barton (✉)
Geography, GIS, and Sustainability, University of Northern Colorado, Greeley, CO, USA
e-mail: karen.barton@unco.edu

S. D. Brunn (ed.), *Transdisciplinary Research Across Knowledge Worlds*,
https://doi.org/10.1007/978-3-032-19491-6_11

International Leader in Climate Policy and the Global Energy Transition

Norway occupies a unique position in the global energy landscape given its reputation as a pioneer in renewable energy and climate action (Eikeland, 2017). The country's abundant natural resources of oil, gas, and hydropower have bolstered its economic development and energy independence. With over 95% of its domestic electricity generated from hydropower, Norway has achieved near-total energy self-sufficiency placing it among the world's most energy-secure nations (Kent, 2014). As such, Norway has played a pivotal role in shaping international climate agreements and environmental policies, becoming one of the first countries to pledge carbon neutrality by 2030, nearly two decades ahead of the European Union's target (Norwegian Ministry of Climate and Environment, 2020, 2023). Norway is ahead of the environmental curve in numerous ways. Financial contributions to global initiatives, such as REDD+ (Reducing Emissions from Deforestation and Forest Degradation), further aid Norway's reputation as a global environmental leader (Broderstad, 2011). However, this leadership is complicated by its role as one of the world's top fossil fuel exporters, a status that has drawn scrutiny for its contribution to global emissions. This paradox highlights contradictions in Norway's energy strategy.

For context, Norway's modern energy economy was revolutionized by the discovery of oil in the North Sea in the late 1960s. This transformative moment propelled the country from a modest economy reliant on fishing and agriculture into a global energy powerhouse. The establishment of Statoil (now Equinor) in 1972 ensured national rather than private control over its oil production. Then in 1990, the creation of the Government Pension Fund Global (GPFG), which contains revenues from oil sales, enabled Norway to secure long-term financial stability and intergenerational equity (Norwegian Ministry of Finance, 2023; Norwegian Ministry of Petroleum and Energy, 2023). Today, the GPFG stands as one of the largest sovereign wealth funds in the world, reflecting Norway's foresight in sustainably managing resource wealth with the prosperity of future generations in mind (Fauchald & Gulbrandsen, 2022).

While oil and gas have been economic cornerstones, it was hydropower that laid the foundation for Norway's energy independence much earlier in the game. Beginning in the early twentieth century, the harnessing of abundant river systems (such as Pasvikelva, Ivala and Teno) enabled widespread electrification in rural communities, powered industries, and fostered financial growth (Midttun & Baumgartner 2019). Jared Diamond's (1997) concept of "geographic luck," the idea that a place's success is determined by its geography, including its climate, natural resources, and animals, is an excellent descriptor for the Norwegian context. Hydropower remains a dominant force in Norway today accounting for 90% of electricity generation and serving as a nearly emissions-free resource (Statkraft, 2021). However, the environmental and social costs of large-scale hydropower projects not just in Norway, but anywhere, cannot be ignored. As we discuss later, developments in Sami territories often proceeded without adequate community consultation or compensation, disrupting ecosystems and diminishing the cultural heritage of Indigenous peoples (Broderstad, 2011; Sami Parliament of Norway, 2023). Addressing these inequities is essential for advancing Norway's renewable energy agenda.

Norway's leadership in the global energy transition, that is, the shift away from fossil fuels like coal, oil and natural gas, is also quite impressive given the country's massive investments in offshore wind (Fig. 1), carbon capture and storage (CCS), and electric mobility (Mind 2005).The Hywind Tampen project, as one example, exemplifies this innovation as the world's first floating wind farm designed to power offshore oil and gas platforms that blends cutting-edge technology with decarbonization efforts. Likewise, Norway's dominance in electric vehicle (EV) adoption, with over 80% of new cars sold in 2022 being electric or hybrid, showcases the success of its incentives and infrastructure (Nilsen & Viken, 2019). Not surprisingly, Tesla leads the market in Norway, accounting for a shocking 20% of vehicles sold in 2024, something that is evident in both rural and urban areas. Norwegians own such a high number of Teslas that social media is replete with jokes about the challenge of finding one's Tesla in parking lots packed with identical cars.

Fig. 1 Fugro's Seawatch Wind Lidar Buoy used to measure wind farm viability. (Credit: Author)

In terms of national accountability, carbon capture and storage (CCS) initiatives such as the Northern Lights project, which are designed to capture CO_2 before it enters the atmosphere, underscore Norway's commitment to addressing its own industrial-based emissions from fossil fuels. This collaboration between Equinor, Shell, and TotalEnergies aims to establish a full-scale CCS infrastructure beneath the North Sea, providing solutions for carbon-intensive industries (Nilsen et al., 2019). However, these achievements still coexist alongside Norway's massive fossil fuel exports exemplifying the paradox of its energy policy. Geopolitics further complicates this energy model. When Russia invaded Ukraine in 2022, it cut off 80 billion cubic meters of pipeline gas supplies to Europe. The cascading effect was that Europe was plunged into an energy crisis. Norway, in turn, increased natural gas exports to Europe, reinforcing its reliance on fossil fuel revenues and complicating its existing green leadership narrative. Indeed, Norway's efforts highlight the complexities of balancing immediate energy security with long-term climate goals in a globally interconnected market (Broderstad, 2011).

Norway's mixed identity, pioneering renewable energy while remaining a major fossil fuel exporter, also raises critical questions about the scalability of its model for other emerging oil rich states. Norway's success story fundamentally hinges on its unique natural advantages, such as its combination of abundant natural resources and oil revenues, a situation that is not universally replicable in places like the United Arab Emirates which only possesses the latter (Nilsen et al., 2019). To align its actions with its climate ambitions and to set an example for sustainable energy transitions worldwide, Norway will need to address its dependence on oil and gas exports. More importantly, it is essential to consider who reaps the benefits and who shoulders the burdens within Norway's energy landscape with particular attention to the experiences of the indigenous Sami people of the North.

"Just Sustainability" and the Ethics of Energy Development

Julian Agyeman's framework of "just sustainability" provides an excellent lens through which to evaluate the convergence of environmental sustainability and social equity as it applies to energy geographies in Norway (Agyeman, 2013). At its simplest, his framework argues that sustainability must be integrated into social justice to ensure equitable distribution of benefits and burdens. In other words, Agyeman asks, "sustainability for whom?" Norway's renewable energy landscape is a prime example of the challenges and ethical dilemmas featured in Agyeman's framework (Barton, 2025). The Sami people, whose ancestral lands extend across Norway, Sweden, Finland, and Russia, currently face systematic marginalization that is exacerbated by renewable energy projects in Norway (Kuokkanen, 2020). While renewable energy is crucial to combating climate change, developments such as wind farms and hydropower installations often disrupt the Sámi 's traditional livelihoods, including reindeer herding and fishing, which are integral to their cultural identity (Eikeland, 2017). Wind farms, with their extensive infrastructure,

Fig. 2 Tromsø: site of the Arctic Frontier meeting and Sami Council. (Photo: Author)

work to fragment grazing lands and disturb reindeer migration routes while hydropower projects alter ecosystems critical to fishing and hunting activities (Nilsen & Viken, 2019). Interviews with reindeer herders in January 2024 confirmed this reality. One herder mentioned how "warmer seasons have affected the timing of migrations and calving, altering our traditional herding calendar" (Personal communication 2024). Another reported: "Wind turbines definitely disrupt reindeer migration routes and grazing areas. The noise, vibrations, and infrastructure create a landscape that reindeer avoid, forcing them into smaller, less suitable grazing grounds. This threatens our traditional way of life and the balance we've maintained with the environment for generations" (Personal communication 2024). In short, renewable energy can potentially disrupt the delicate balance between herding practices and environmental conditions that were honed over multiple generations and when coupled with climate change these impacts can be profound (Fig. 2).

In addition to direct impacts from renewable projects, the decision-making processes surrounding renewable energy frequently, but not always, marginalize Sámi voices. This reflects Norway's broader exclusion patterns and the way in which this reinforces historical injustices (Nilsen et al., 2019). Situated in the far north of Norway, the Sámi's physical distance from Oslo also contributes to their spatial marginalization. This type of exclusion contradicts the participatory ideals central to both sustainability and Indigenous rights frameworks (UNDRIP, 2007). Without inclusive governance models, projects risk perpetuating inequities and failing to adhere to the principles of "just sustainability" (Agyeman, 2013). As a result, the Sámi have had to advocate for their rights in both national and international forums such as Arctic Frontiers in Tromsø which highlighted the need for new and better governance that prioritizes ethical and equitable practices (Kuokkanen, 2020).

Sami Folklore and Geographies of Environmental Stewardship

It is important to underscore the unique role that Sámi culture plays within Arctic landscapes. Sámi traditions trace back thousands of years, making them one of the oldest continuous Indigenous cultures in Europe (Minde, 2005). Archaeological evidence suggests that the Sámi people have inhabited the Sápmi region, which spans northern Norway, Sweden, Finland, and the Kola Peninsula in Russia, for at least 5000 to 7000 years (Fig. 3). Their way of life, deeply connected to reindeer herding, fishing, and hunting, has adapted to the Arctic and sub-Arctic environments over millennia (Mikkelsen, 2022). Oral histories, traditional knowledge, and spiritual practices reflect a profound relationship with the natural world rooted in both animism and shamanism. My visit to the Sápmi region of Norway in January

Fig. 3 Sami land as recognized by international law. (Map: Shutterstock)

2024, during Hurricane Ingunn's landfall, highlighted the profound challenges the Sámi face in adapting their livelihoods to increasingly extreme climatic conditions. Witnessing how the shifting targets of climate change impact their way of life made the experience deeply impactful and sobering.

Sámi people possess a rich tradition of folklore that is intimately connected to their environment serving as both a cultural archive and a roadmap for future actions (Minde, 2005). They have proven to be excellent stewards of the natural environment, not only in Norway, but also what is now Finland and Sweden and parts of Russia. Sámi Sami oral histories and legends emphasize sacred landscapes, such as *sieidi* (sacred stones or sites), which symbolize the interconnectedness of humans and nature (Nilsen & Viken, 2019). For example, a prominent *sieidi* in the Alta region of northern Norway, often visited by Sami herders, was historically adorned with offerings like reindeer antlers to ensure the prosperity of their herds. Also, Noitatunturi, located in Pyhä-Luosto National Park, is a sacred mountain revered in Sámi shamanistic traditions and mythology. The site, marked by *sieidi* stones, served as a spiritual center for rituals connecting the Sámi people to the natural and supernatural realms. Such sacred sites and the rituals associated with them underscore the deep respect that defines the Sami's relationship with their land. These narratives not only foster a reciprocal relationship with the environment, they also illustrate principles of ecological balance integral to reindeer herding and fishing conservation (Sara et al., 2021).

Storytelling among the Sámi functions as a form of environmental education, offering insights into seasonal changes, migration patterns, and ecosystem dynamics (Bull, 2020). Tales of *stallo,* a mythical giant often convey moral lessons about the dangers of overexploitation and ecological imbalance which ultimately promote sustainable behaviors (Kuokkanen, 2020). In the face of climate change, these narratives have gained renewed relevance and importance across northern Norway. Rapid environmental shifts in Sápmi, the Sámi homeland, threaten traditional ways of life and Sami cultural leaders are revitalizing folklore to highlight these challenges and advocate for sustainability (Lechner et al., 2019).The integration of Sami folklore with contemporary climate strategies underscores the value of traditional ecological knowledge in addressing global challenges (Bjørklund, 2019). For instance, their cyclic grazing practices for reindeer herding align closely with modern conservation principles exemplifying ecological balance and resilience (Sara et al., 2021). Thus, collaborations between Sámi communities and policymakers, blending traditional wisdom with scientific approaches, can offer pathways to effective Arctic preservation not just for indigenous peoples but for all concerned (Turi, 2020).

Agyeman's "just sustainability" framework can help us to better conceptualize the intersection of Norway's renewable energy policies and Sámi rights. Achieving *just* sustainability requires meaningful participation and fair compensation for the Sámi, ensuring that energy development aligns with social justice principles (Agyeman, 2013). By integrating Sámi knowledge and prioritizing their rights, Norway can lead by example demonstrating how renewable energy transitions can be both environmentally sustainable and socially equitable. In fact, Norway is already a leader in cross-cultural humanitarian efforts that are demonstrated by the success of Jan Egeland's team at the Norwegian Refugee Council; the nation simply needs to build on these existing achievements. We now turn to case studies that delve into the interplay between Norway's energy economy and the rights of the Sami people by highlighting the challenges and opportunities in achieving equitable sustainability.

Case Study 1: The Fosen Vind Project and Green Colonialism

The Fosen Vind project, situated in the Trøndelag region of Norway, serves as a key example of the tensions that occur between renewable energy development and the rights of Indigenous peoples (Fig. 4). As Europe's largest onshore wind farm, the Fosen project includes six wind farms and over 250 turbines, generating enough electricity to power hundreds of thousands of homes (Statkraft, 2021). While the wind farm aligns with Norway's renewable energy targets and its commitment to reducing greenhouse gas emissions (Norwegian Ministry of Climate and Environment, 2020), it has also become a flashpoint for controversy due to its impact on the Sámi community, whose traditional lands and livelihoods are directly affected by their placement.

The Sámi have expressed concerns about the encroachment of industrial projects of any form on their ancestral territories for some time (UN Permanent Forum on Indigenous Issues, 2021) but the Fosen project is of particular concern to the Sámi. What is unique about Fosen is its massive scale relative to other developments, with over 1057 megawatts of installed capacity (Bjørnarå & Nybø, 2021). The 278

Fig. 4 Fosen Wind Farm. (Photo: author)

turbines in the Fosen project were placed on land critical to reindeer grazing, essentially disrupting migratory routes and fragmenting habitats essential to the health and sustainability of reindeer herds (Nilsen et al., 2019). Beyond these practical and everyday use challenges, the Sámi have argued that the project infringes on their cultural rights and has violated their sovereignty over their traditional lands (UNDRIP, 2007). Thus, the Sámi's resistance was not just about their reindeer livelihood, but about their ability to control traditional and accustomed territories as they have done for millennia.

In a landmark 2021 ruling, the Supreme Court of Norway found that the construction of the Fosen Vind turbines violated the Sami people's rights under international law (Broderstad, 2011). The ruling specifically referenced the United Nations Declaration on the Rights of Indigenous Peoples (UNDRIP, 2007). In short, the court emphasized that the project failed to adequately consider the Sámi's cultural and economic dependence on the land and concluded that the turbines represented a significant breach of their rights (Norwegian Supreme Court, 2021). This decision was hailed as a major victory for indigenous rights advocates by setting a precedent for holding large-scale renewable energy projects accountable to international standards (Amnesty International, 2021). However, despite this legal victory, the turbines at Fosen Vind remain operational today, exposing a significant gap between legal recognition of indigenous rights and their practical enforcement (Nilsen et al., 2019). Economic interests, including the substantial investment in renewable energy infrastructure and Norway's role as a global leader in sustainable energy, have taken precedence over the court's ruling. This disconnect underscores the difficulties in implementing legal frameworks that protect indigenous communities when they conflict with powerful economic and political forces (UN Human Rights Council, 2022) even in a nation as progressive as Norway.

The Fosen case has drawn widespread attention to the concept of "green colonialism," a term used to describe the ways in which the global push for renewable energy can replicate colonial patterns of exploitation (Bull, 2020). Green colonialism occurs when indigenous lands are appropriated for environmental initiatives without adequate consultation, consent, or benefit-sharing (Sara et al., 2021; Whyte, 2018). In the case of Fosen, this chapter argues that the wind project certainly exemplifies a form of green colonialism as the Sámi community has been sidelined in decision-making processes and left to bear the environmental and cultural costs of Norway's energy transition (Broderstad, 2011; Nilsen & Viken, 2019).

Looking ahead, the Fosen Vind project highlights the urgent need for more inclusive and equitable approaches to renewable energy development. Meaningful engagement with indigenous communities, respect for their traditional knowledge, and adherence to international legal standards are essential for avoiding the pitfalls of (Agyeman, 2013; Lechner et al., 2019). Furthermore, the case underscores the broader ethical dilemma facing the global renewable energy movement, viz., how to reconcile the imperative of combating climate change with the equally critical need to protect human rights and cultural diversity (IPCC, 2022). By addressing these challenges, Norway has the opportunity to demonstrate green, socially just leadership in crafting sustainable energy policies that uphold the principles of equity. The

Fosen Vind case serves as a powerful reminder that achieving a truly just energy transition requires more than technological innovation; it demands a commitment to social and cultural inclusivity at every stage of the process (Agyeman, 2013; Whyte, 2018).

Case Study 2: Hydropower Development and Sami Resistance

Hydropower, widely regarded as the backbone of Norway's renewable energy strategy, also plays a critical role in the country's transition to sustainable energy. With over 90% of Norway's electricity generated from hydropower, its significance cannot be overstated (Statkraft, 2021). However, this reliance has also become a flashpoint for conflict particularly with the Sámi people. The Alta Dam controversy of the late 1970s and early 1980s exemplifies the tension between state-led development and Indigenous rights. The Alta River project, recently immortalized in the 2023 film, "Let the River Flow," was envisioned as a means to increase Norway's renewable energy capacity. Yet it sparked one of the most significant environmental and Indigenous rights protests in Norwegian history. In short, the Norwegian government's decision to build a large dam on the Alta River led to widespread resistance from the Sámi and environmental activists. Central to the opposition was the argument that the dam would disrupt reindeer migration patterns, flood Sámi lands, and undermine the livelihoods of the Sámi people (Broderstad, 2011). These concerns resonated with broader environmental movements around the globe during that time period, aligning Indigenous rights with global ecological sustainability.

The protests escalated in intensity with dramatic acts of resistance including hunger strikes outside the Norwegian Parliament and physical blockades at dam construction sites (Fig. 5). These efforts drew international attention, placing Norway under scrutiny for its treatment of Indigenous peoples (Bjørklund, 2019). Despite the global outcry, the Norwegian government proceeded with the dam's construction, citing the national importance of hydropower for economic and energy security. Although the Alta Dam was ultimately built, the good news is that conflict served as a watershed moment for the Sámi rights movement. Surprisingly, the controversy catalyzed significant legal and political reforms. For example, in 1989, the Sami Parliament (*Sámediggi*) was established, providing a formal mechanism for the Sami to influence policies affecting their territories and cultural heritage. In addition, Norway adopted international conventions recognizing Indigenous rights, such as International Labour Organization (ILO) Convention 169 in 1990, further institutionalizing protections for the Sami (Kuokkanen, 2020).

The legacy of the Alta controversy continues to influence contemporary debates over hydropower development in Sámi territories. While hydropower remains central to Norway's energy strategy, newer projects are now more carefully scrutinized for their social and environmental impacts. Recent developments, such as proposals for wind and hydropower installations, have reignited concerns about land encroachment and the prioritization of national interests over Indigenous rights. These

Fig. 5 Protest over Alta Dam. (Photo: Courtesy of Alta Museum, Norway)

ongoing tensions highlight the need for more inclusive decision-making processes that balance energy development with the rights and well-being of Indigenous communities (Eikeland, 2017).

At its core, the Alta Dam case underscores the complex intersection of renewable energy development, environmental justice, and Indigenous sovereignty. It serves as a reminder that sustainability must encompass not only ecological considerations but also the equitable treatment of all communities affected by development projects. By integrating Sámi voices into the planning and implementation of renewable energy initiatives, Norway has the opportunity to model a more just and inclusive approach to sustainability as it considers its energy *future.* By better incorporating the needs and livelihoods of the Sámi, Norway can also work to break down the barriers that create the parallel universes Stan Brunn delineates here in this volume. Some of that work begins with a reconceptualization of Norway's "sovereign wealth fund," a major byproduct of its oil economy.

Norway's Oil Fund, the Sámi, and Ethical Investment

Norway's sovereign wealth fund, formally known as the Government Pension Fund Global (GPFG), is among the largest and most influential investment funds in the world (Fig. 6). Valued at over $1.7 trillion (NOK and USD) as of 2025, the GPFG plays a critical role in managing revenues generated from the country's oil and gas

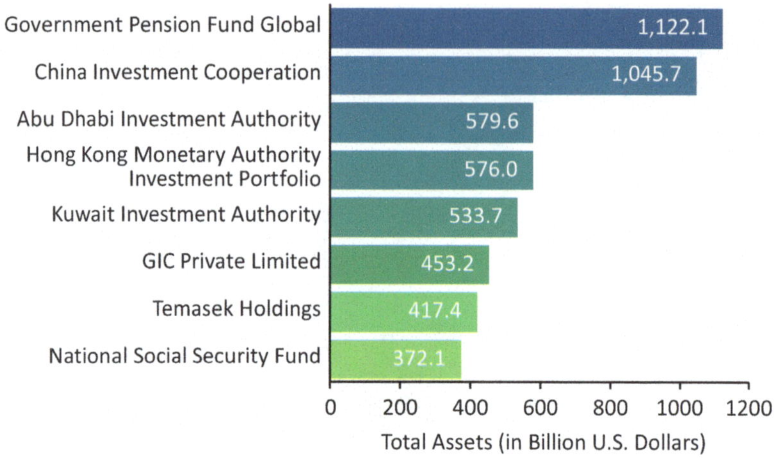

Fig. 6 Norway Oil Fund (Government Pension Fund on the graph) is the largest of the World's biggest sovereign wealth funds. (Data source: Statista 2021; Graph: Author)

exports (Norges Bank Investment Management, 2024). The sheer scale of the Fund, as it is known colloquially, is so impressive that the Fund's value can be tracked via a publicly accessible website with kroner amounts in real time (Nilsen et al., 2019). The Fund is widely recognized for its leadership in ethical investment, maintaining strict exclusion criteria for companies involved in controversial activities such as coal mining, deforestation, tobacco production, and human rights abuses (UNDRIP, 2007). These exclusions are guided by ethical guidelines established by Norway's Ministry of Finance, emphasizing sustainability and accountability (Nystuen, 2011). Our Fulbright team was afforded the opportunity to speak with the Ministry of Finance in Oslo in 2022, and public officials were genuinely open to answering questions about how investments are decided and who benefits most. This degree of transparency is emblematic of Norway, a nation that consistently ranks highly on international corruption indices for its integrity.

Despite its progressive approach to investment, the GPFG has faced significant criticism for the paradox at its core: its financial foundation rests on revenues derived from fossil fuels, a major contributor to global climate change. This contradiction has led critics to describe Norway's sovereign wealth fund as an example of "climate hypocrisy," wherein a country simultaneously positions itself as a global leader in ethical investments while profiting from the fossil fuel industry. This critique is particularly salient in light of Norway's international commitments to climate action, such as its adherence to the Paris Agreement and its ambitious domestic goals to achieve carbon neutrality by 2030 (Norwegian Ministry of Climate and Environment, 2021). Norway's paradox is not unlike a firefighter fueling the flames—working tirelessly to extinguish the global fire of climate change while simultaneously profiting from selling the oil that feeds it.

Sámi leaders have also articulated their concerns regarding the GPFG's ethical framework and its potential to address injustices faced by Indigenous communities.

In many cases, they have done this in person. Sámi communities (and other citizens) are permitted to peacefully protest to the King by going to the palace with a written petition, a symbolic act rooted in the historic tradition of seeking redress directly from rulers during their reign. This approach evokes the longstanding Norwegian practice where subjects could appeal to their sovereigns as a means of ensuring justice and addressing grievances under the rule of law. My fieldwork interviews with Sami villagers in Tromsø reveal a shared belief that the Fund could, and should, take a more active role in promoting social and environmental justice. For example, by channeling investments toward Sami-led renewable energy initiatives or cultural preservation projects, Sami communities argue that GPFG could align its operations more closely with principles of just sustainability. Such measures would not only mitigate the environmental impact of energy development in Sami territories but also empower Indigenous communities to play a leadership role in the green transition (Broderstad, 2011).

The GPFG has certainly begun to take incremental steps toward incorporating Indigenous perspectives into its investment strategy. The Council on Ethics, which advises the fund, has now (broadly) included "human rights considerations in its recommendations, although Sámi -specific measures remain underdeveloped" (Norges Bank Investment Management, 2024). Expanding these measures to include targeted investments in Sámi -led projects could serve as a model for how sovereign wealth funds worldwide can reconcile ethical principles with economic realities. Ultimately, the future of the GPFG lies in its ability to balance Norway's financial interests with its commitments to ethical and sustainable development. Transforming the fund into a vehicle for systemic change, that is, one that not only divests from harmful industries but one that invests in solutions led by Sámi communities, could redefine the role of sovereign wealth funds in addressing global inequalities. This approach would help Norway bridge the gap between its climate leadership and the realities of its resource-dependent economy by demonstrating a path toward a more inclusive, sustainable future (Kent, 2014). Ironically, Norway leads the way in humanitarianism through its strong commitment to international aid, peacebuilding, and human rights advocacy. The country consistently ranks among the top donors to humanitarian causes relative to its GDP, funding critical programs that address global challenges such as poverty, education, and refugee support (Kuokkanen, 2020). Norway has also played a key role in mediating peace processes worldwide, including in Colombia, Sudan, and the Middle East, by demonstrating a proactive approach to conflict resolution. Scaling these issues downward to aid and recognize Sámi communities would further strengthen the country's international reputation for promoting human rights and serve as a win-win solution.

Toward a Restorative Energy Policy: Bridging Parallel Universes

We argue in this chapter that the spatial tensions surrounding wind farms, hydropower, and fossil fuel exports in Norway highlight the urgent need for a new, restorative and geographically informed approach to energy policy. These conflicts underscore the interplay between human–environmental systems where global priorities for renewable energy transition collide with the local cultural and ecological landscapes of the Sámi. This dynamic represents a convergence of parallel universes: one driven by the spatial imperatives of climate mitigation and the other rooted in preserving cultural identities among Arctic peoples. Restorative energy policy can help reconcile these dimensions, emphasizing equity, inclusion, and the rights of marginalized communities to ensure that the benefits of renewable energy projects are shared while minimizing their environmental costs.

Integrating Indigenous knowledge into energy policy is the first step in addressing these challenges. The Sámi people, whose cultural and spatial practices span across Norway, Sweden, Finland, and Russia, embody traditional ecological knowledge deeply tied to Arctic landscapes. Their practices, such as reindeer herding and migratory patterns, are spatially adaptive strategies that balance human activities with fragile ecosystems. Highlight how such localized, place-based knowledge offers critical insights for achieving sustainable development. By embedding this knowledge within energy policy, planners can create projects that are both environmentally restorative and culturally respectful.

First, reforming the consultation processes is a vital next step toward bridging the spatial divide between energy infrastructure and Indigenous territories. Norway's current consultation frameworks often fail to adhere to the principles of free, prior, and informed consent, leading to human-environmental conflicts (Øverland, 2021). Turi (2020) emphasize the importance of participatory governance that engages Indigenous stakeholders throughout all phases of energy projects. Establishing more joint decision-making bodies with equitable representation could provide a model for spatially inclusive and just energy planning. These processes must consider the spatial footprints of energy infrastructure and their impacts on human–environmental interactions.

Second, the role of Norway's sovereign wealth fund represents another area for restorative justice. As the world's largest sovereign wealth fund, it holds the capacity to reallocate resources toward projects that prioritize spatial justice and environmental sustainability. For example, funding *community-led* renewable energy initiatives can empower local populations to have spatial control over energy production and distribution. Turi (2020) demonstrates that such initiatives often result in greater public acceptance and stronger connections between communities and their environments. Beyond energy infrastructure, the sovereign wealth fund could support reparative programs aimed at revitalizing Sami cultural landscapes, such as initiatives to restore traditional grazing areas or promote cultural mapping projects.

Restorative energy policy recognizes that renewable energy development and cultural preservation must come together in shared spaces where communities like the Sami and the environment intersect (Ellingsen, 2024). As climate change accelerates, it puts greater pressure on resolving conflicts between these priorities. Norway's unique landscape, which includes Sámi territories, positions the country as a global leader in finding this balance. This approach aligns with global efforts like the International Labour Organization's Just Transition, which emphasizes fairness and inclusivity in sustainability. By integrating Sámi knowledge, reforming consultation processes to include Indigenous voices, and supporting projects that benefit all communities, Norway can set a powerful example of how to bridge the gap between cultural heritage and environmental progress during energy transitions.

Conclusion: Lessons for Global Energy Governance

Norway's energy paradox offers valuable lessons for the global energy transition, especially when viewed through the lens of the distinct experiences of the Sámi people and mainstream Norwegian society. These groups often live in what feel like parallel universes, shaped by fundamentally different relationships to land, climate, and resources. While the Sámi have stewarded fragile Arctic ecosystems for centuries, their ways of life are increasingly threatened by global climate change and the encroachment of renewable energy projects that risk repeating the patterns of green colonialism.

For Norway to truly lead in sustainable development, it must acknowledge the importance of its Indigenous peoples in the far north and integrate their voices and needs into national energy policy. This includes protecting their cultural heritage, recognizing their traditional knowledge, and addressing the disproportionate impacts they face (United Nations, 2007). Norway has already demonstrated remarkable leadership in balancing economic growth with environmental stewardship, and building on this foundation by prioritizing the rights and well-being of the Sámi people can set a global standard for equitable and inclusive climate action. The ecosystems in which the Sámi live are both longstanding and incredibly fragile making it vital to ensure that climate solutions do not come at the expense of Indigenous rights or environmental integrity. During my second visit to the far north in January 2024, I witnessed firsthand the effects of a hurricane Ingunn, an explosive storm that reached the northern extent of Tromsø, making it the most damaging storm to hit Scandinavia in 30 years. The Sami and other northern communities are experiencing the impacts of climate change in real time and a sense of urgency is understood by both sides of this parallel universe.

The lessons from Norway also extend well beyond its borders, offering a critical reminder for policymakers, activists, and researchers: the global energy transition cannot be truly sustainable without equity and inclusion. Avoiding green colonialism and fostering collaboration with Indigenous communities should be at the forefront of international efforts. Ultimately, achieving a just and sustainable energy future requires not only technological innovation but also a strong commitment to letting all voices be heard, and then addressing these concerns through meaningful, tangible action. This involves translating dialogue into policies and initiatives that genuinely reflect the needs and aspirations of all stakeholders, ensuring that progress is both inclusive and transformative. While there have been challenges as outlined above, the Norwegian government and the Sámi are ultimately on the same side by both striving for sustainable futures that respect the land and its people. As climate change accelerates, it puts greater pressure on resolving conflicts between these priorities.

References

Agyeman, J. (2013). *Introducing just sustainabilities: Policy, planning, and practice*. Zed Books.

Amnesty International. (2021). *Norway must respect Sami rights in Fosen case*. https://www.saamicouncil.net/news-archive/report-uncovers-human-rights-violations.

Barton, K. (2025). Exploring geography and education for sustainable development through issue-based case studies. Chapter 15 in S. W. Bednarz, J. Mitchell (Eds.), *Handbook of geography education, Springer International Handbooks of Human Geography*.

Bjørklund, I. (2019). The Sami and development projects in Norway: Reindeer husbandry and wind power. *Indigenous Rights Quarterly, 14*(2), 45–63.

Bjørnarå, H. T., & Nybø, S. (2021). Sami land rights and renewable energy: Lessons from the Fosen case. *Arctic Review on Law and Politics, 12*(1), 34–50.

Broderstad, E. G. (2011). The promises and challenges of indigenous self-determination: The Sami case. *International Journal, 66*(4), 893–907.

Bull, K. S. (2020). Green colonialism and the Sami: Ethical dilemmas in renewable energy development. *Journal of Indigenous Policy Studies, 15*(3), 22–39.

Diamond, J. (1997). *Guns, germs, and steel: The fates of human societies*. W. W. Norton & Company.

Eikeland, P. O. (2017). Hydropower and indigenous rights: The case of the Sami in Norway. *Energy Policy, 39*(7), 49–63. https://doi.org/10.1016/j.enpol.2017.10.045

Ellingsen, H. (2024). The temporary-political dimensions of green colonialism through wind power development at Fosen, Norway. *Human Arenas*. https://doi.org/10.1007/s42087-024-00450-1

Fauchald, O. K., & Gulbrandsen, L. H. (2022). The Norwegian sovereign wealth fund: Ethical challenges and global implications. *Global Environmental Politics, 22*(1), 77–101.

IPCC. (2022). *Climate change and land*. https://www.ipcc.ch/srccl/

Kent, N. (2014). *Sami peoples of the north: A social and cultural history*. Oxford University Press.

Kuokkanen, R. (2020). *Reshaping the university: Responsibility, indigenous epistemes, and the logic of the gift*. UBC Press.

Lechner, A. M., McIntyre, N. & Witt, K. (2019). Green colonialism: The hidden costs of renewable energy in indigenous territories. *Renewable Energy Journal*.

Mikkelsen, A. (2022). Reindeer herding in the age of renewables: Sami perspectives on land use conflicts. *Geographical Review, 112*(2), 145–167.

Minde, H. (2005). Assimilation of the Sami—Implementation and consequences. *Acta Borealia, 22*(2), 121–136.

Nilsen, H., Sjafjell, B., & Richardson, B. (2019). The Norwegian Government Pension Fund Global. *Risk Based Versus Ethical Investment, 88*(1), 65–78. https://www.econstor.eu/bitstream/10419/215469/1/10_3790_vjh_88_1_065.pdf

Nilsen, T., & Viken, A. (2019). Tokenism and power: Sami perspectives on environmental governance in Norway. *Environment and Society, 10*(1), 29–48.

Norges Bank Investment Management. (2024). *Government pension fund global annual report.*
Norwegian Ministry of Climate and Environment. (2020). *Norway's climate action plan.* https://www.regjeringen.no/contentassets/a78ecf5ad2344fa5ae4a39441 2ef8975/en-gb/pdfs/stm202020210013000engpdfs.pdf
Norwegian Ministry of Climate and Environment. (2021). *Norway's climate action plan.*
Norwegian Ministry of Climate and Environment. (2023). *Norway's climate goals and actions.*
Norwegian Ministry of Finance. (2023). *The Government Pension Fund Global: Background and purpose.*
Norwegian Ministry of Petroleum and Energy. (2023). *Norway's oil and gas history.*
Norwegian Supreme Court. (2021). *Judgment in Fosen Vind case.*
Øverland, E. (2021). Indigenous governance and Norway's energy transition: Bridging environmental and cultural concerns. *Nordic Policy Studies, 28*(3), 89–105.
Sami Parliament of Norway. (2023). *Hydropower development and Sami rights.* https://sametinget.no/_f/p1/id38eedb4-9f07-400e-8bcc-5515ab706ec1/sami-parliament-report-to-ilo-on-c-169.pdf
Sara, S., Rasmussen, T., and Krovel, R. (2021). *Defining the limitations and opportunities in consultation with the Sami: The cases of the Arctic railway and the Davvi vindpark.* https://arcticyearbook.com/images/yearbook/2021/Scholarly-Papers/17_AY2021_Sara.pdf
Statkraft. (2021). *Fosen Vind: Europe's largest onshore wind power project.* https://www.statkraft.com/globalassets/0/.com/0-start-page/2021-annual-report_sustainability-chapter.pdf
Turi, J. (2020). Energy colonialism in the Arctic: The case of wind farms on Sami lands. *Nordic Journal of Human Rights, 38*(4), 320–339.
UN Human Rights Council. (2022). *Report on indigenous rights in the context of climate action.*
UN Permanent Forum on Indigenous Issues. (2021). *Challenges facing indigenous communities.*
UNDRIP. (2007). *United Nations declaration on the rights of indigenous peoples.* UN General Assembly.
United Nations. (2007). *United Nations declaration on the rights of indigenous peoples.* UN General Assembly.
Whyte, K. P. (2018). Indigenous science (fiction) for the Anthropocene; ancestral dystopias and fantasies of climate change crises. *Environment and Planning: Nature and Space, 3*(1), 224–242.

Indigenous Geography and the World Wide Web

R. D. K. Herman

It was 1999 and I was working at the Smithsonian's National Museum of the American Indian (NMAI) in Washington, D.C. Quite by accident I stumbled across a proposed project called Geografía Indígena, or "Indigenous Geography." This ambitious, but at that point vaguely defined, project was a major step in presenting Indigenous community-driven cultural geography and ontology using the World Wide Web.

The idea derived from two consultations NMAI had undertaken with Native educators in July 1998 and August 1999. Community Services asked these educators, "If we could do one project for you, what would it look like?" Well, it needed to engage fully with Native culture and language; it needed to foster traditional values; it needed to expand on the relationship between Native communities and their environments; and it needed to do so while embracing computer technologies, so that Native youth would see that traditional culture and modern technologies can go together.

Geografia Indígena was envisioned as a three-pronged program that would serve to: (1) produce a web-based portfolio of indigenous knowledge about Indian communities in the Western Hemisphere; (2) produce educational materials, curricula and professional associations for teaching Geography in accordance with national geography standards but as tailored to the needs, values, and perspectives of Native America; and (3) develop internet and multimedia technological capacity within Indian communities that would, among other things, allow them to preserve their own cultural information and to expand on their community web sites.

Thus, it would provide a mechanism for both meeting the need to perpetuate Native culture, and for technologically empowering Native communities to further this process themselves. This "knowledge/technology transfer" was intended to operate in a way that was culturally appropriate and useful. Technology was seen as

R. D. K. Herman (✉)
The Pacific Worlds Institute, Hilo, HI, USA

S. D. Brunn (ed.), *Transdisciplinary Research Across Knowledge Worlds*,
https://doi.org/10.1007/978-3-032-19491-6_12

the tool for achieving the goals of cultural preservation, while the content—coming from the communities themselves—was understood as the driving force behind the project. In addition to that innovation, it presented cultural geography from the perspective of Indigenous peoples.

I was fresh from completing my doctoral degree at the University of Hawai'i, where I had been studying Hawaiian language, culture, history, and environments. I had also been thinking about Indigenous epistemology and ontology, so as a geographer I introduced myself to the project leader, Tim Johnson (Mohawk). Using some grant money, Tim brought me on board as a consultant and assigned me the task of figuring out how to actually do what they were proposing. Tim himself came from a journalism background, and he roped in his friend Jose Barreiro (Taino), who had been publisher of the journal *Native Americas*, to assist.

In the Fall of 1999, Tim sent me and Jose, along with another Community Services member, Susan Secakuku (Hopi), down to the village of Sipawlovi, Hopi Second Mesa, to interview her father Ferrell Secakuku, an elder of that village. I had never done ethnographic work before, but Jose—a journalist—was an experienced interviewer, particularly in Indigenous settings, and I learned by observation. We were there for just 3 days, and I tried to focus on getting information to fill all the boxes in the grid I had developed.

Becoming a Web Developer

Geografía Indígena proposed a long list of topics to cover, including a list of words to be presented in the Indigenous languages. While working for Tim, I took a night course on website architecture at George Washington University and set about mapping out the different topics they wanted to cover. I came up with eight categories which, while laid out below as a linear diagram, I conceptualized as a circle—the Medicine Wheel—moving from "Origins" through the range of human-environmental knowledge and ending (or beginning, since it's a circle) with "Ritual."

Now began the process of taking raw oral data and turning it into a website. After our return from Hopi, I transcribed the tapes myself and began sorting the information into MS Word files for each topic. Since the information came almost entirely from one person, there was not much need to edit other than to clean up the "ums" and "ahs" and move some pieces around to go in the right thematic box. Once that was done, I scoured the internet for additional information, including maps, weather and climate data, demographic and economic indicators, and more. There was not a lot out there.

Tim had sent an NMAI photographer to the village before we had done the interviews. A very talented photographer, she did not understand the project and could not have had any idea what we were specifically looking for. And until we had done the interviews, neither did we. So, she got some landscapes and textures and a range of other photos that, however beautiful, did not really suit the bill. I did not have access to use those photographs.

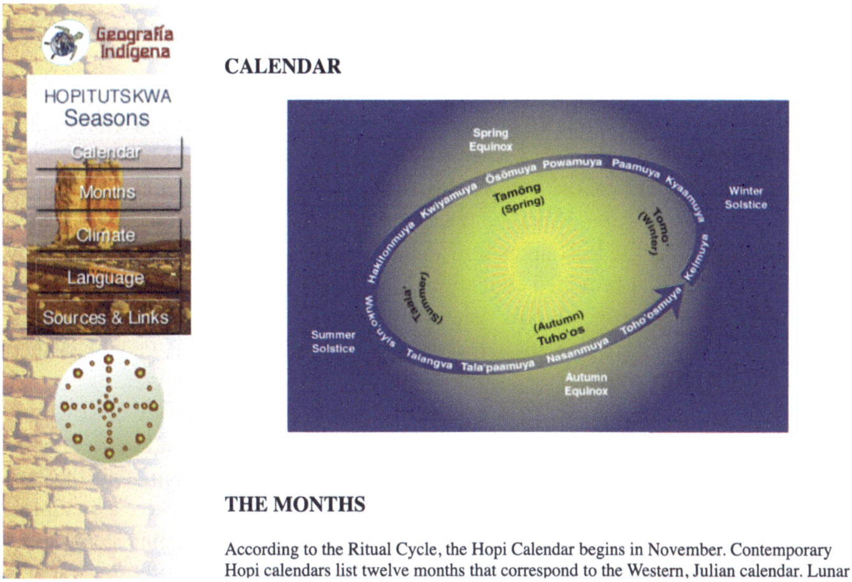

Fig. 1 Indigenous geography website mock-up

I figured out how to use MS Excel to turn climate data into climographs. And I began teaching myself Adobe Photoshop and Adobe Illustrator. NMAI would undoubtedly have their own graphics team do any necessary work, but I had ideas about how I wanted to visually display certain kinds of information. Most importantly, I taught myself HTML, which was fairly uncomplicated at that time.

I was not technically the website producer—NMAI had professional staff for that. But I was eager to see a product, and NMAI was moving very slowly. So, I produced a version of the website using frames, a now mostly deceased style that split the screen into separate panels: one used for navigation, the other for content (Fig. 1).

That Fall (1999), NMAI was contacted by the Administration for Native Americans (ANA), another federal agency. They wanted to fund an NMAI project, and Geografía Indígena was chosen. I wrote the proposal for $160,000 to begin doing 24 communities over three years. It boggles my mind now to think of how outrageously ambitious that timeline was.

Then over the Christmas holidays, Tim suddenly left NMAI. Meanwhile a congressional spending freeze meant that ANA could not give us the money. I left the museum and, without a decent job prospect, set about to do my own version of Geografía Indígena for Pacific Islands. It was, after all, intended to provide a replicable model. So, I replicated. Fortunately, that Fall I got a tenure-track job at Towson University, where I carried out much of the work.

Pacific Worlds

I drafted a new grid for Pacific Islands (I hadn't actually been to any besides a few Hawaiian Islands) and shopped it around in Honolulu. The Hawai'i Geographic Alliance agreed to help run teacher workshops once a pilot site had been done. My old department connected me with Carlos Andrade, an older Hawaiian working on a PhD focusing on Hā'ena, Kaua'i, a place at the end of the road on the island at the end of the chain. Pacific Resources for Education and Learning (PREL), a federally funded non-profit that provided educational resources to the American Pacific (former territories in Micronesia plus American Samoa) was interested but said their indirect cost rates were too high for most grants. The Bernice Pauahi Bishop Museum was not interested.

I called the project Pacific Worlds. Pacific Islanders share a legacy of voyaging and exploration out of Asia over thousands of years and inhabit small tropical environments that have a number of similarities. But the cultural forms and the languages and discourses that transmit them vary in key ways that make each cultural ontology its own "world." By linking these community-derived ethno-geographic portfolios together in a consistent framework, the similarities and differences across the region can be explicated.

I registered the domain name pacificworlds.com and, at the appropriate time, found a website hosting company to take it on. I got a seed grant of $2000 from the Hawai'i Council for the Humanities and went with Carlos to Hā'ena, where we interviewed three local residents (one being himself). I did all the photography—on film. Digital cameras were in their very early days. I did not own one, and they were very expensive (in 2001 I bought a Fuji Finepix with 3 megapixels maximum resolution, for $849, but it also had 8x optical zoom and an attachable wide-angle lens, both extremely useful for my purposes). I also requested permission for use of existing maps, photos and graphics where possible. Because this was all non-profit, non-commercial, and educational, permission was usually granted.

I completed and launched the website that Fall (2000). I then informed the various folks back in Hawai'i that I had a product. Most were not interested, but PREL got back to me immediately. One of their federally-funded projects had failed to deliver, and they needed a deliverable. "How much to buy into this?" They asked. Stupidly, I told them—to the penny—what my out-of-pocket expenses had been that fell outside the seed-grant funds. I did not account for the hundred or so hours I had put in. But this began a new partnership. Over the next few years, PREL threw $100,000 dollars at the project to create four sites in Micronesia.

I meanwhile applied for large grants from the National Endowment for the Humanities (NEH). The first time, I got fairly positive reviews, but not good enough. I reapplied the following year, now having two communities online and extended material. The reviews got worse. And they were even worse the third year, when I had even more and better content. What happened?

It was clear from the reviews that they had come to the "politically correct" position that this work should be done by the Indigenous islanders themselves. Except that they were not doing it—only I was. It also became clear that the NEH committee members were Humanities people for whom "fieldwork" meant time in an archive or museum somewhere, not out in an actual field. "We like to give money to scholars, not to airlines," the committee chair told me over the phone.

Mapping out Indigenous Geography for the Internet

I was my own boss now, so with Pacific Worlds, I was able to draw on what had been proposed for Geografía Indígena, but also to reimagine the whole project in ways that might better serve Pacific Islands. I stuck with the basic format of eight themes or chapters but had one focusing on the land and another on the sea. And because my previous research had focused on the changes wrought by colonization and missionary influence, I added historical layers to bring the content up to the present. Cultural geography in the islands today cannot focus on the traditional culture alone but needs to account for the changes wrought over the past few centuries that put them where they are today.

The grid that I proposed before going to Hāʻena in 2000 got tweaked a little as I took on new sites outside of the Hawaiian Islands. Figure 2 shows the grid that resulted.

Several years later, after working with island educators, it became clear that certain topics were missing. In particular, I was told that food was an essential part of island culture that needed elaboration. I added *Health* to that, stemming from my own engagement with traditional healing techniques, because here again—as with so many other facets of this project—traditional knowledge presents a very different and more holistic approach to Being. So, for the last two websites I did—and eventually, to retrofit the ones already done—I expanded to a ten-section format. The environmentally focused portions of *Land* and *Sea* were combined into one chapter called *Living World*, while the food-focused parts of those chapters went into a new *Sustenance* chapter (Fig. 3).

In either case, there is an introductory or home-page section (I think of it as the "departure lounge") where you are introduced to the place, where it is, the *Location* of the village or land division within the larger political entity, and an *Orientation* to major features, place names and sites. *Getting There* provides a sense of relative location. For example, to get to Ulithi Atoll from anywhere outside of Micronesia, one flies first to Guam, then to Yap, and finally on a very small plane to Ulithi Atoll, where a rusted pickup truck (one of three vehicles on the island Falalop) picks you up and takes you to the one small hotel. Or you can come by the one ship that makes monthly runs across Micronesia. What it takes to get there gives that sense of relative location.

Critical is the self-introduction of the *Guides* whose words form the bulk of the text. Each self-introduction comes with a photo. In the spirit of NMAI's notion that "the community are the experts," this provides the position of each speaker and their authority to speak.

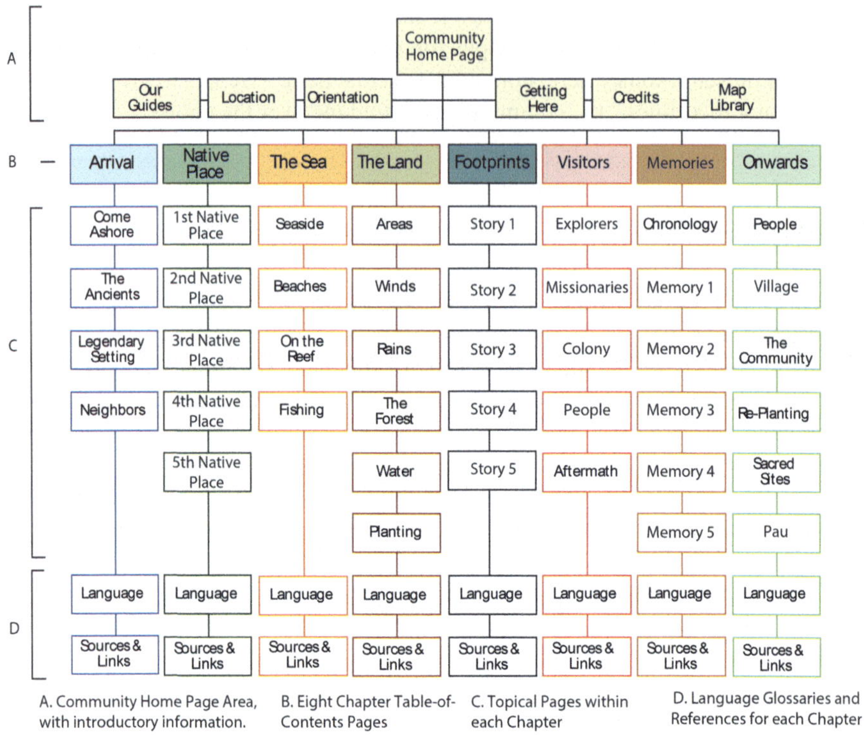

Fig. 2 Eight-theme sitemap

Now comes the mapping out of the community's indigenous geography by themes. These themes ("chapters" of the website) are somewhat chronological and sometimes not. Past and present mingle together in places and separate in others.

It begins with *Arrival*. This actually refers to *their* arrival—traditions of voyaging (if available) or mythical origins, *Legendary Setting*, the creation of the island(s), who the *Neighbors* are (more relative location), and a description of how this village or land division is geographically defined. In Palau, for example, villages have stone-lined walkways that end in platforms, and these platforms mark the end of the village. Here in the past there may have been young men patrolling as guards to protect the village.

Next is *Native Place*. This focuses on five cultural sites that relay important information about place-based culture. They will differ from place to place. In the case of Guam and Saipan, where Spanish colonization eradicated traditional settlement patterns, sacred sites, and a great deal of place-based knowledge, the communities talked about the church, cemeteries, and cultural events such as weddings and fiestas in which, underneath the Spanish overlay, aspects of the traditional practices remain.

Living world speaks of the physical environment in cultural understandings. *Areas* explores how the different environmental zones are categorized in the Native

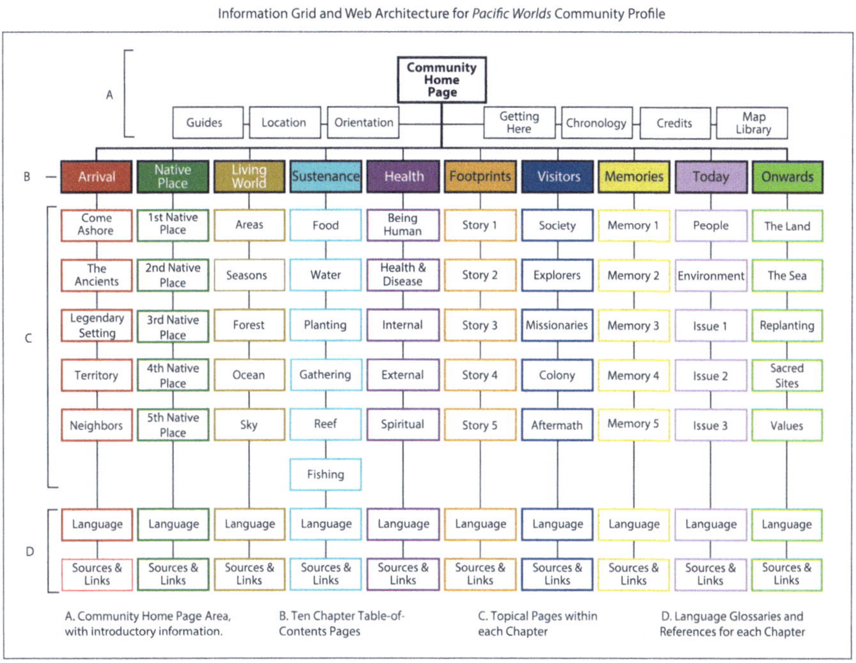

Fig. 3 Ten-themed information grid and web architecture for Pacific Worlds community profile

terminology and conceptualization. *Seasons* looks at the traditional calendar, around which so many activities such as fishing and farming are organized. Then *Forest*, *Ocean*, and *Sky* go into finer details about these realms and their inhabitants.

Sustenance describes how food is acquired out of this environment, starting with traditional foods and methods of preparation, then what the elders remember eating (usually a hybrid of traditional and later introductions). *Water* discusses freshwater sources, which is hugely important on small islands. Then come *Planting*, *Gathering* (both plants and more recently the hunting introduced animals), and—moving out into the sea—what is gathered from the *Reef* and obtained by *Fishing*. What species are important to this village? What technologies, protocols, and traditions are involved in their planting/gathering/preparation? Here in particular we often find a blend of traditional and modern practices and technologies, but the traditional values always come through.

Health follows closely from *Sustenance*. It starts with the question, What does it mean to be human? And once we understand that, how do we understand *Health and Disease*? And once we understand that, how do we treat disease? Three categories of treatment are discussed: *Internal Treatments* (things ingested, such as herbal medicine,) *External Treatments* such as massage and steam baths, and then the psychological and spiritual aspects of healing. In Indigenous cultures, health is often understood as communal. It is an inter-relational phenomenon. This is similar to the Ayurveda and other holistic approaches to medicine.

Footprints is like *Native Place*, only where *Native Place* looked at cultural (i.e., human-made) sites, *Footprints* looks at naturally formed legendary sites: physical shapes in the landscape that have stories. These stories, like so much of mythology, convey important cultural values about proper behavior and cultural values.

Now things get more strictly chronological. *Visitors* recounts the colonial era: what was *Society* like before these outsiders arrived? Who were the *Explorers* that arrived here, and what happened? *Missionaries* almost invariably followed, and some form of *Colony* ensued. *Aftermath* leads into the twentieth century.

Next, *Memories* looks at significant historical events, periods, or practices that the elders remember. In Micronesia, a lot of it had to do with the Second World War. But there were other things as well, such as Carabao racing on Guam, tsunamis, Americanization.

That brings us up to *Today*. *People* looks at changing demographics, and the impact on culture-environment relations. *Environment* looks at the state of the ecosystem today. Then there are three current issues that the community members feel are most important, from invasive species to food sovereignty, and in Hawai'i, political sovereignty.

The final chapter is *Onwards*. By this point we have, in GIS-speak, seen many layers of human-environment knowledge and traditions, from the ancient times to the present, from the traditional culture at the time of contact to the situation today. The question is, where we going from here? How do we want to manage the *Land Today*? The *Sea Today*? The *Sacred Sites*? What are we doing in terms of *Replanting* the practices and traditions that helped us survive in the past, back into the land and sea, and back into future generations? This discussion ends with *Values*: what are those values we want to see perpetuated? Particularly given climate change and its impact on ecosystems, the same plants and animals may no longer be available, but the practices and principles regarding ecosystem management and social fabric can be carried forward.

In most Indigenous languages, the future is *behind* us. We are looking at the past—what is *be-fore*—to guide us in what we do *after*. The navigator on a voyaging canoe sets the initial course by looking back at the island disappearing over the horizon. So too should the values honed by generations of living on small islands help those peoples move forward today, despite the forces of modernity.

Cartography

I am not much of a GIS person. When I was in graduate school, GIS was just starting to emerge. We all had to learn this new word "pixel." I had taken a computer cartography course that used pre-GIS programs. So, I contracted others to produce GIS maps for me. Towson University has a Center for GIS that was associated with the Geography Department, so I hired their students to do the work for me. Today I use QGIS on a Macintosh to do my own work, and Hawai'i has a wonderful online GIS Data Center that provides the data. In Micronesia I was given GIS data on CD

ROM. I took some historical maps and rendered them in Adobe Illustrator so that, as with a GIS map, I could turn certain layers on or off to illustrate specific topics (Fig. 4).

In addition to a limited use of GIS, I also used screenshots of GoogleMaps and GoogleEarth (once these came into existence), including 3-D images on occasion, and put labels on them using Adobe Illustrator or Photoshop. Historical maps were employed where appropriate, sometimes with added color rendering to show

Fig. 4 Kramer Map

specific features, such as the streams and *'auwai* (agricultural waterways) of Nu'uanu Valley.

For every community site, I created a Map Library page where users can look up the different maps used in that site, clicking either an icon to open the image itself, or a link to take them to the page on which it appears.

Then there is the matter of locating sacred sites or sensitive locations. Here I tended to be rather general in putting points on maps, and never provided enough location that one could readily find these sites. This is to protect them from disturbance, though in some cases the sites are well known. Nu'uanu valley is the area upland from Honolulu, and for this area I turned to Sterling and Summer's (1978) *Sites of O'ahu*, a compendium of fragments of text from manuscripts in the collection of the Bernice Pauahi Bishop Museum. Nu'uanu turned out to have a surprising number of storied places, and I endeavored to find each one of them if possible. Some had been destroyed, but many had not. One storied spring had actually been turned into a park, with a bronze plaque telling the story. Another pool in Nu'uanu stream I finally located using a map from 1920. It too was part of a park—a children's playground that is fenced and locked. I had to hike up the stream to it.

For the Nu'uanu's *Footprints* chapter, the last content page is *Other Storied Places*. For this I created a map of the valley using a GIS hillshade image, with clickable points that open pop-up windows with whatever information—and ideally an image or two—was available for that site.

Technical Details

In the early 2000s, web design had grown from its infancy to the toddler stage. Monitors were mostly still small (600x400 pixels), and dial-up internet was still common. There was not a lot of web-weaving software available at that time, and my seed-grant money wasn't going to cover it. So, I wrote the original website by hand in HTML using Text Edit on a Macintosh. I could then load the page in a browser (Netscape, at the time) to see how it turned out, then go back in and make changes to the code as necessary.

Beyond simple HTML, there are three fancier tricks I used to make the website look and operate the way I intended. First is the use of mouseovers. Mouseover is a simple Javascript code inserted into the header section of the web page, identifying "active" and "inactive" images. The default is the inactive image, but putting the mouse over it changes it to the active image. No knowledge of Javascript necessary: it was easy to copy the template and insert my own information. You create two images that have the same prefix—"turt," for "turtle," in this example—plus a suffix of either "on" or "off." "Turton" is designated as the image to show up if the mouse comes over (the "active" image), and "turtoff "when it is "inactive." And then, in the code that places the image on the page, the mouseover action is designated as follows: onMouseOver = "act('turt')" onMouseOut = "inact('turt')". The turtle logo in the banner for each page actually flaps its flippers to take you to the Home page. You can check out how this works on the original Hāʻena site: https://www.pacific-worlds.com/haena/index.cfm.

At that time, drop-down menus were at best in their infancy and were created using Macromedia Flash, which would have required mastering another software program (I am a *cultural* geographer; my interest in software goes only so far). They were not yet the prominent technique that they later became, and I did not use them. So, in a manner that is now old-fashioned, each page shows, below its button bar, a list of the pages in that chapter, with the active page's font in a different color. Similarly, the active-chapter button appears different from the other chapter buttons.

I repeated this structure at the bottom of each page with the icon for that chapter along with a set of links to the homepage for that community, the Map Library and Site Map for that community, and the Pacific Worlds Homepage. The sponsors who supported the production of that community site are also identified by text and icons with links to their websites.

I did use a Cascading Style Sheet (CSS). This is actually a simple text document that sets styles for fonts of different classes, much like you can set Heading styles in MS Word. The code sets the size, weight, font family, and even font color, and can set underline, bold, etc. Here is the code for the class I called "subtitle":

```
.subtitle {.
    font-size: 12pt;
    font-weight: 700;
    font-family: Arial, Helvetica, sans-serif;
    color: #990000;
    }
```

You can apply the class to an entire table cell—<TD CLASS="topnav">, "TD" being the code to designate a table cell—and then within that, specify/unspecify specific fonts as needed for headings or whatever. Because I was writing the site by hand in HTML, I would have to type in the class I wanted to use, e.g., <font class = subhead>. When I later adopted Adobe (originally Macromedia) Dreamweaver as my web-writing software, applying classes was easily done with a click. The beauty of the CSS (Cascading Style Sheets) is the ease of providing consistency in appearance across the website. I applied different font colors to links that had already been visited versus those that had not. Once I had more than just the one community, I worked to give each different community its own style. Hence each community has its own CSS.

The last bit of "fancy" code that I did involved creating the drop-down menu that allows the user to jump from a page in one community to the same page in another community. This required using Adobe ColdFusion, a program that can do a lot of things if you can be bothered to learn it. I needed just this one piece, and I had to get outside help for the code, which sits in a separate text document. Then I just stick one line of code in the right place on the web page to pull it in: <cfinclude

template = "../../jumpmenu.cfm">. That was the fairly easy part. It also requires that instead of the suffix ".htm," the file names end in ".cfm."

That jumpmenu code engages a database that identifies each community, each chapter, and each page. And that database was written in Microsoft Access, the most inaccessible software program I have ever tried to use. It does not even come with the MS Office package anymore. I had to get a programmer to do it for me, and to this day, I need a programmer to make any changes to it. I do not know why, it is a very simple database, but Access has some secret ingredient that makes it work where Excel does not, and it is extremely difficult to use where Excel is easy.

Design Elements

I used tables to create a consistent shape, size and appearance, setting the width at 600 pixels, again because at that time (2000), that was the size of a lot of monitors. I also kept the images at fairly low resolution, because internet speeds were still pretty slow for those not on a T1 connection. Dial-up internet was still common. I recall using an internet café on Palau in 2001 and waiting 5 or 10 min for one simple jpeg to load.

I used Photoshop to create a banner for all the pages in each community site (Fig. 5). I strove to use images representing the inland area and the shore. I then split that graphic into three sections at 300, 150, and 150 pixel widths. These pieces went into three table cells, and two of them had mouseovers that changed them to show they were active links—one to take you to the community homepage, and one to take you to the Pacific Worlds homepage. Then at the bottom of each page I put a footer: an ocean view (either out from the shore, or back at it), rounded to suggest the hull of a canoe. These environmental cues help emphasize that this is a geography project.

I created eight buttons for the different "chapters" to go below the banner. I also created icons for the eight themes. For Hawai'i, I used images of petroglyphs. For other communities, there were few if no petroglyphs to work from, so I sought representative graphics or produced ones of my own to use for these icons (Palau has extensive storyboard art). They appear on each page of each chapter in each community website.

Again, the way I had conceived of this project was that it would use a consistent grid of topics for each community. So, within the overall website, one would click on a map on the home page to go to the home page for a specific community. There, a sort of sub-website with its own style would open up. But the architecture and navigation structure for each of these sub-websites would be the same. There are pros and cons to this. Some argued to me that each community should be able to set its own structure to what information they want to share and how they want to share it. I understand that position. But as an educational tool—it is the educational angle more than the cultural documentation angle that brought in the grant funding—the consistent structure allows for comparative study. So, from any given thematic page

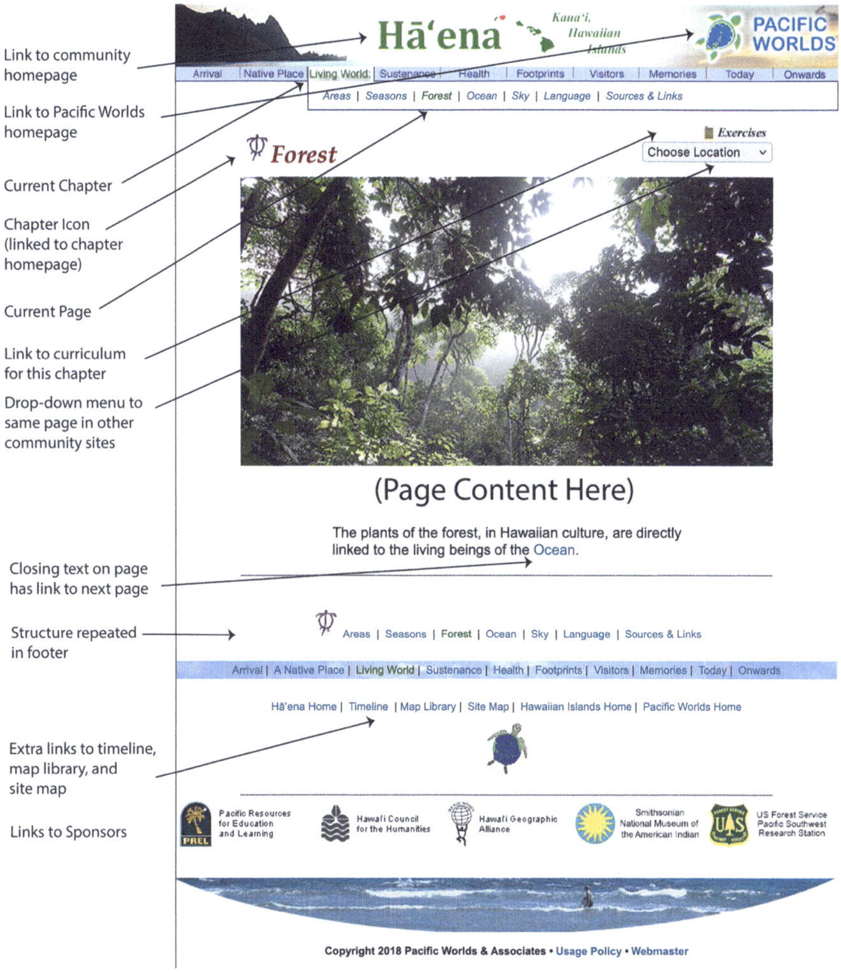

Fig. 5 Sample page from the Hā'ena site, showing the various design and navigation components

in one community—*Planting*, for example—the user could go to a drop-down menu and jump to exactly that same page in another community. A teacher could thus use the site to have students explore just one topic deeply across all the communities in the project. This compare-and-contrast approach is integral to the purpose of Pacific Worlds: How are these "worlds" similar and different? What practices, understandings and values are shared or unique? What terminology is used in each case to talk about the world they live in?

Each chapter also has a home page, a table of contents, with images and a short description of the content pages. On that chapter homepage is also a link to a short essay entitled "About This Chapter." These explain what this website is trying to accomplish in terms of conveying "Indigenous Geography."

In the second and third grants I received from the Hawai'i Council for the Humanities, they asked that I include scholarly essays on some related topic. This is a great idea, and I included three: one on Voyaging, by Carlos Andrade, who was a sailor with the Polynesian Voyaging Society, and two on the art of conducting Ethnographic interviews. One of these is "Methodology for an Oral History Project," from Brian Calliou (Sucker Creek First Nation) (2004), excerpted with permission from an excellent published article. The other I commissioned: "Doing Oral History with Native Hawaiians," by Charles Langlas, University of Hawai'i at Hilo. Links to these essays are placed in the appropriate thematic areas of the community websites and provide an additional resource.

Fieldwork

In each community that I worked on, I had a guide—a community member to make introductions, take me around, and otherwise point me in directions I needed to go to find out certain things. For the four communities I did in Micronesia, PREL had offices there which worked with local authorities (for example, someone from the Department/Ministry of Historic Preservation, the Department/Ministry of Education, and the local museum), to decide which village would be the focus. Once that was determined, they then found a specific person to be my guide. In the Hawaiian Islands, I started with Carlos Andrade, then had other connections for Nu'uanu, Kawaihae, and He'eia. In the last case, I connected with the local Hawaiian Civic Club, as well as three Hawaiian non-profit organizations working in that land division.

Interviews are the first step in the data-gathering process. These are open-ended discussions. I do have in my mind the grid of topics I want to cover, and my guides take me to specific people for specific topics. I learned early on that people will talk about what they want to talk about, and you need to roll with it. It is not a matter of having a list of questions, and often they will bring up things I never thought of and want to pursue. The whole time, I am picturing my grid and mentally saying, "Ah, that story goes over there, this one goes over there." Interviews were recorded and later transcribed.

The information obtained often directed me to certain sites in the area. This leads to the second and equally important piece of the process: walking the land. This includes not just the normal roads and pathways, but sometimes off the beaten trail, as well as into the sea. I visited caves with petroglyphs, old, abandoned sites of ceremony or habitation, and even offshore islets and underwater reefs that were storied places. For the latter I had to purchase underwater camera equipment. I started with a cheap disposable on Palau, then an underwater case for a small Nikon, and finally a GoPro as the technology improved.

Walking the land is as much about getting a feel for the place as it is for seeing sites and taking photographs. It is for experiencing all the sensory input that a place has to offer. Eating the local foods goes with that, and in some places, chewing betel nut and drinking tuba (coconut toddy). And in some cases, walking the land led to the encounter with sites of significance that the locals took for granted until I asked about them.

The local museum—if there is one—is another important source. Here, one may find cultural and historical artifacts that illustrate the stories told by the elders. In some places, I was able to obtain copies of maps from the local planning office, or posters of wildlife from the local environmental office. The Historic Preservation offices often had old photographs I could use—again, all of this was non-profit and for educational purposes. And in Hawai'i, the State Archives has historical documents, images, and maps.

Back home, I would search bibliographic and online resources. The Trust Territory of the Pacific Islands (TTPI) was a United Nations entity formed of the former Japanese territories in Micronesia and was administered by the United States from 1947 to 1994. Its photography archive is online and proved an amazing source of images.

Finally, some Western style information is brought in to round out the science aspects in particular, and to provide a point of discussion between traditional and Western perspectives. Statistical and demographic data provide greater detail on phenomena (climate, population and so forth) that is useful to supplement the oral records, as well as providing a demographic history of the area that relates to other changes in land, society and economy.

When I first began this project, I did not read anything in advance of my visits to these new places. This was more by accident than intent but was effective in that I went with no pre-conceived notions from the Western literature. I think this is crucial—at least when I first began—because the whole intent of the project is to gather and present Indigenous perspectives. The published literature is often written by Western anthropologists or historians, who frame things in Western terms and concepts, and had I read that in advance, it may have pre-framed my perspectives before I hit the ground. Instead, it came fresh from the mouths of community members, on site.

Processing the Data

Processing the materials gathered involves seeing what you have, determining where the different bits of information fit within the information template (the site map), preparing the materials for use, and finally composing the disparate pieces into well-crafted web pages. Doing this often requires gathering more data—either text or images—to connect things or fill gaps. Figure 6 outlines the overall method for the data processing.

Step one is transcribing the interviews. I would then make a copy of those "raw" files and clean them up—removing all the "ums" and "ers" and rendering more complete and cohesive sentences, while sticking as close as possible to the actual words used by the contributor. Then, I would make a copy of that file and use it for

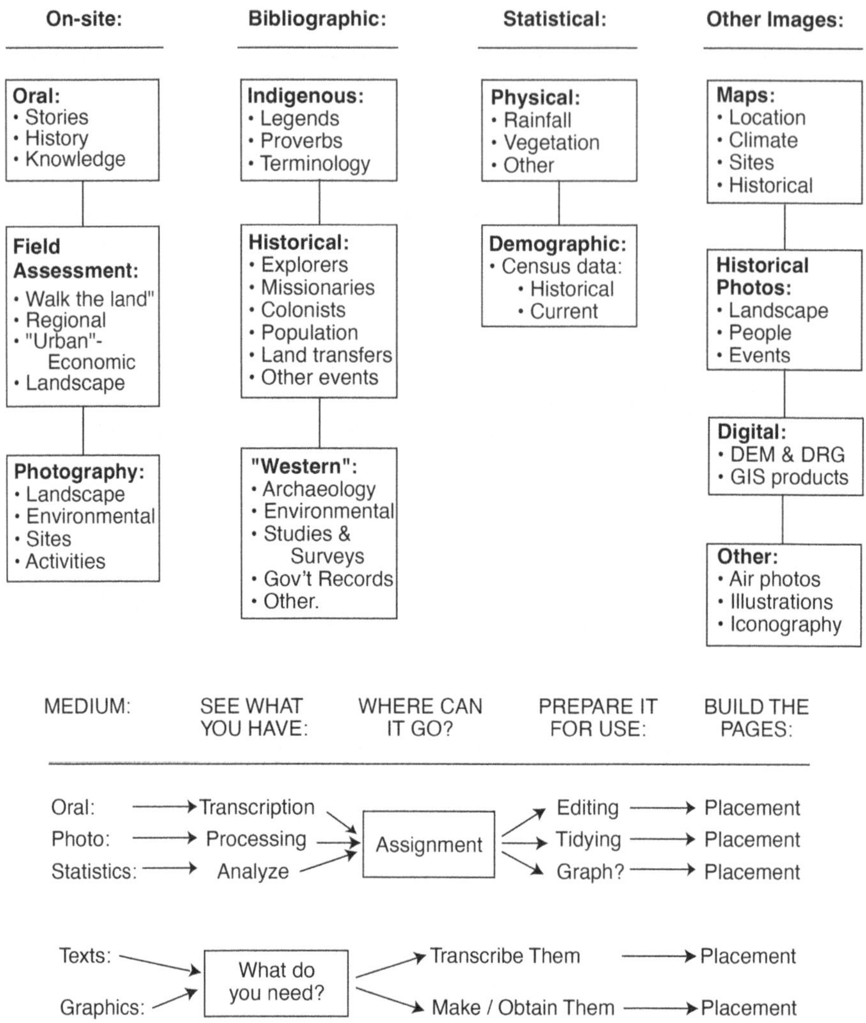

Fig. 6 Data processing

a massive cut-and-paste job. I would create blank Word files for each box in the information grid and copy and paste the relevant passages from the interviews into those files.

Next, I would take each of these documents and organize and craft the different pieces from different contributors into a cohesive narrative. Each piece is a direct quote and the speaker is identified. Sometimes I had to add a little narration to glue the pieces together, but I tried to keep that to a minimum. Where appropriate, I would add bibliographic data to supplement the community voices.

It is at this point that I invariably found some gaps in the material. Sometimes I could fill those creatively: "OMG, William Yaad told me this awesome story that I didn't see where it fit, but now I see I can use it *here*." But a second, and sometimes a third, research trip was always needed to fill all the boxes in the grid. Occasionally I would find published material to fit the bill and acquire permission to reproduce it.

On the other hand, sometimes I had material that did not fit any of the boxes or elaborated on a small topic or site in some detail. Having gathered this oral information, I was determined to use it, so in these cases I would create a separate pop-up page for the information. It would be linked from the appropriate page in the grid. For example, on Ulithi Atoll I was told the entire process for harvesting coconut sap and making tuba, a fermented beverage also known as coconut toddy. I linked to this from the *Planting* page: "Click here to learn about the making of tuba." For some communities, there are few or none of these, and for others, a great many. They are listed on the Site Map for each community.

Then, there is processing the photographs: creating folders for the different topics (not quite as finely grained as the information grid itself, because one never knows exactly where each image might be used) and organizing and labeling them. I generally did this each night when I was in the field, so that they would be backed up on my computer (once I had gone digital).

Finally, I created graphics as necessary to represent certain information. One such graphic is the calendar—a pretty complex Illustrator production that took a great deal of time. Figure 7 shows the calendar for Palau.

The result is that each page is a blend of images and text, the text being a narrative usually woven from different voices, each identified. My favorite example is the "Legendary Setting" page of the Yap-Ulithi website. Here, the legend of Liomarer and the creation of the atoll is told by six different contributors. It reads as though the contributors are sitting around a table telling it together, when in fact they were separate and distinct interviews in which each added pieces of the story.

I try to place images (photographs, maps, graphics, object images) frequently enough that there is almost never a huge unbroken block of text. Still, the pages can be long. I am not opposed to reading. I grew up with actual books. I once met a web designer from PREL at an airport bar somewhere in the region, and he explained that he wanted his websites to feel like a video game, with bells and whistles. I take the opposite approach: just read it. The content speaks for itself.

Curriculum

This is an educational project, and towards that end I produced a curriculum to go with the website. This curriculum follows the chapter structure, presented questions and exercises based on the topics in each chapter. In some cases, these ask the students to compare and contrast practices across different communities. In other cases, it asks that they collect the similar information for their own community.

I examined the national educational standards as well as what existed for geography standards at the time. The Hawai'i Department of Education had a fourth-grade Hawaiiana requirement and a seventh-grade Pacific Island studies requirement,

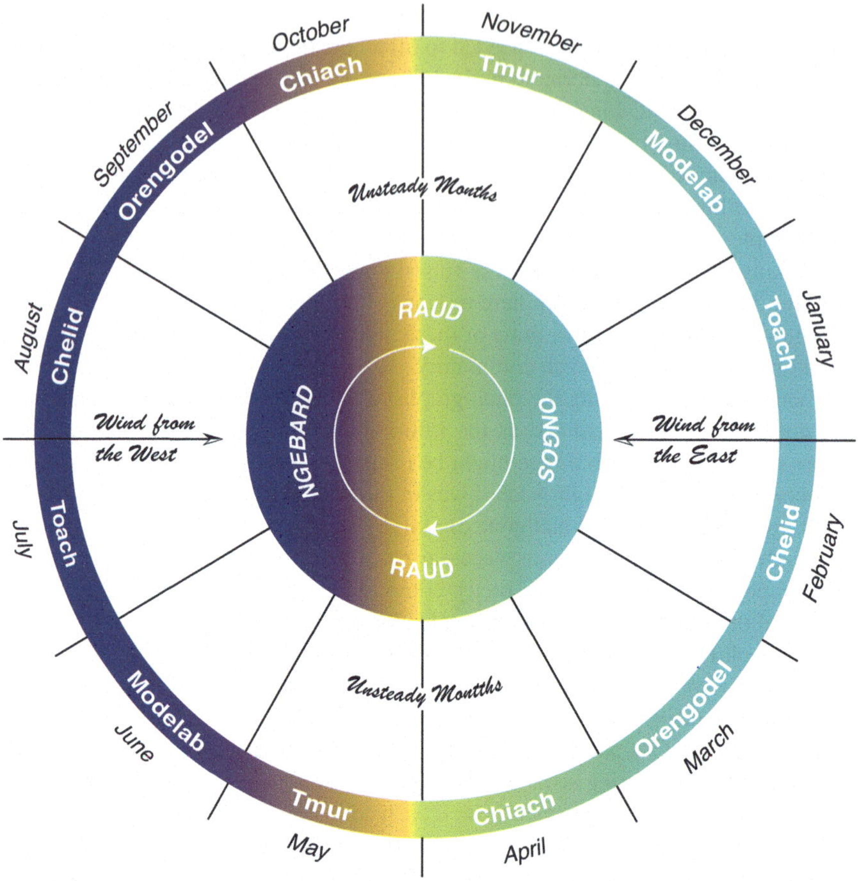

Fig. 7 Palau Calendar

so these were the main targets of the curriculum. I would like to have hired a K-12 curriculum specialist, but in the end wrote it myself.

I also held over a dozen teacher workshops to introduce educators to the project and have them go through it and learn how to use it in the classroom. Many of these were held in the Hawaiian Islands, but some were held at PREL's annual education conferences in Micronesia. Funding from the University of Hawai‘i's Center for Pacific Island Studies allowed me to travel to Fiji and Samoa in the South Pacific. (For many, "South Pacific" is a generic term for all Pacific Islands, but Hawai‘i and most of Micronesia are actually north of the Equator.)

The Pacific Worlds website has an entire education sub-section that includes the curriculum in various formats, information about the teacher workshops, and some discussion of how the project addresses U.S. educational standards. It offers the curriculum in MS Word format and encourages teachers to butcher and modify it to suit their needs.

There are also links to the exercises in each page of each chapter, that go to an online version of the curriculum.

The Academic Angle

Towson is the State of Maryland's number two university by a considerable gap. It had previously been the state teacher's college. The teaching load at that time was 3/3, and the expectations for tenure below that of a research-one university. Not every professor had refereed publications, with some focusing more on the teaching and service areas. Non-refereed publications were welcome. Now the question was, where does a website fit in to this? By the time I came up for tenure, I had completed seven community projects. I counted these as separate publications—they came out at different times, had their own home pages and styles, and so on.

There was no denying that these were major bodies of work. The later ones ran to well over 100,000 words. They involved fieldwork (a geography staple) and other data collection. And they were all grant funded, which brought money in to the university—always a huge plus. But a website was still a new animal. I have no doubt that the grant money was a major factor: I had brought in about $125,000, which for a social-science/humanities project is a lot. And I also had six refereed journal articles, a couple of book chapters, and an enormous track record of organizing paper and panel sessions for conferences of the Association of American Geographers. And, of course, presenting at these and other conferences as well.

But I have wondered: had I been at a research-one university, how would these websites have been viewed? Were they peer reviewed? Yes, in the sense that they were reviewed by members of the community (NMAI folks held that: "We believe the community are the experts."), but not in the academic sense. This was a whole new way of presenting geography, both in content and in medium. *Content* in that the data is almost entirely oral material from community members, curated into thematic essays. *Medium* in that these are content-rich websites. So where would they fit into tenure considerations? I do not know that there is an answer.

The Value of Indigenous Geography

I have elsewhere reflected on what I deem to be the importance of indigenous geography and how it compares and contrasts with Western geographical approaches (Herman, 2008) and refer the reader to that essay for an extended discussion. The point I would like to make here is the importance of destabilizing the edifice of Western rational thought as a monolithic and hegemonic ontology. We scholars are used to the relatively dry, fact-checked, peer-reviewed, removed and objectified world of Western academic writing. This includes geography textbooks, where views of others around the world are framed into Western approaches (such as the "development" model) and present environmental exploitation for individual profit as normal behavior—the legacy of the *homo-economicus* model.

Media such as Pacific Worlds and Geografía Indígena (now offline, alas) put the cultural perspectives of the peoples themselves first: "This is how we understand this and how we act accordingly." And for most Indigenous peoples—and Pacific Islanders in particular—the emphasis is more on the health of the community as a whole, and the land and sea on which it depends for survival. Pacific Worlds does not present in an omniscient, third-person narrator's voice, but in the voices of the peoples themselves.

I experienced first-hand that this can be an eye-opener for students. In a Geography of Pacific Islands class, I gave students a few introductory paragraphs from anthropologist William Lessa's (1966) booklet on Ulithi. Here is a sample:

> Canoe houses, better referred to as sheds, line the shore and provide not only shelter for their canoes but a clubhouse for the men….Friends often visit them and join in the gossiping and lounging. Some men may be found performing a light chore, such as twisting coconut husk into twine or carving out a loom sword. Children like to dart in and out but may sit still long enough to listen to the conversation and yarns of the men.

When they had read it, I asked, "What do you think?" They all shrugged and nodded, expressing that it was fine, it all made sense, nothing wrong with it. I asked, "Is anyone in this description doing anything *real*?" Now that they were thinking about it, they gradually recognized the paternalistic viewpoint of this white visitor describing brown people as not really adults with agency, but more like cartoon cut-outs. But at first, they thought it was okay. This is where Pacific Worlds, in presenting the cultural geography as told by insiders, differs substantially:

> The major purpose of the Men's House," Chief Taithau explains, "is that whatever comes to the island by sea, whenever a boat comes to this island, there must be someone from the boat to come to the Men's House—directly to the Men's House. They must come from the boat to the Men's House to report the purpose of their visit. If they don't do that, then the boat is automatically placed on what was called *ubud*. That is, we confiscate the boat. (Taithau, 2003)

I then assigned students to review certain parts of the Palau-Airai website, starting with the *Guides* page, and then to answer a few questions. One student came up to me the next class and said, "I don't get it. Who's speaking? The English is not very good." They were used to academic writing with that objective, scholarly tone. I said, "Did you read the *Guides* page?" "No, I skipped that." "Well," I said, "go back and read that first, and then the other material." She came to me the next day very excited: "OMG, it's the chief of the village that's speaking!" This reflects how the material presented in this project differs from, and can augment, academic writing and teaching.

Time for a Do-Over

Web-developing has changed a lot in the past 25 years. First there are hardware changes. Monitors have gotten much bigger. Smart phones and tablets have come into being and require that web design be "responsive" to whatever device calls them up. Internet connections have gotten a lot faster. And the entire World Wide Web is much more robust than it was before.

Then there are the software changes. Dynamic HTML (DHTML) was developed in the late 1990s and was coming into vogue around the time I was building *Pacific Worlds*, but I did not learn of it until later. Dynamic HTML combines HTML, CSS, and JavaScript to allow for interactive web pages, animations, menus, forms, and more. Extensible Markup Language (XML) is a data-management tool that allows different applications to communicate with each other over the internet and define the structure and format of data exchanged between different applications. Asynchronous JavaScript and XML (AJAX) is a combination of web technologies that makes web applications more responsive to user interaction. In short, there is a lot of coding to learn, and that is not where I want to spend my time. And who knows where things will be in another 25 years?

When in 2021 I created a website for my non-profit, the Pacific Worlds Institute, I used Wordpress for Business. This has all those dynamic advantages, but I get frustrated using their templates and not being able to manipulate the code to do exactly what I want. I am not nearly as pleased with the site as I was with creating the *Pacific Worlds* website itself. That said, I would like to migrate all the existing material into a new form that meets current standards. I just do not have the time and money to do it right now. The entire site is over 1000 separate pages.

More importantly, I hope to train community members themselves to do the work in documenting their own place-based cultural geographies. Especially as the older generation passes away, and in some cases, as the islands themselves sink beneath a rising sea, now is the time to capture that knowledge and wisdom.

And finally, to give the communities themselves (though defining who that is can be tricky) more control over their websites, so that they may update and make changes as they see fit.

For the overall purpose of Pacific Worlds, aside from documenting these Island cultures, is presenting the values of sustainability that is so integral to these traditions. An island in the middle of the sea is a model planet earth in the middle of space: what you have there is all you have. You need to make it last. You need a system of values and practices that enable you to live together in the finite space, treating each other, the living world, and the island and ocean themselves with respect. In contrast to the "development" model that often underlies contemporary geography teaching—positing the exploitation of resources for profit as normal and natural—these cultures remind us of the respect and humility we need to have if we are going to survive on Island Earth.

View the website at www.pacificworlds.com.

References

Calliou, B. (2004). Methodology for recording oral histories in the aboriginal community. *Native Studies Review, 15*(1), 73–105.

Herman, R. D. K. (2008). Reflections on the importance of indigenous geography. *American Indian Culture and Research Journal, 32*(3), 73–88.

Lessa, W. A. (1966). Ulithi: A micronesian design for living. In *Case studies in cultural anthropology*. Holt, Rinehart & Winston.

Sterling, E. P., & Summers, C. C. (1978). *Sites of O'ahu*. Bishop Museum Press.

Taithau, C. A. (2003). *Men's houses*. Pacific Worlds Yap-Ulithi. Pacific Worlds. http://pacificworlds.com/yap/native/native2.cfm

Parallels and Intersections: Patriotism and Pandering on U.S. Commemorative Postage Stamps Since 1957

Thomas L. Bell

> Designs in connection with postage stamps and coinage may be described, I think, as the silent ambassadors on national taste. William Butler Yeats, Speech at Irish Free State Senate on the Coinage Bill, March 3, 1926

Introduction

This quote by the famous Irish poet and statesman W. B. Yeats, perfectly encapsulates the purpose of this chapter, viz., how do postage stamps convey a sense of nationalistic pride? To find the answer I draw on my own experience as a long-time stamp collector. I started collecting stamps in 1957 as a member of an after-school stamp club organized by my junior high school social studies teacher. Though interested in stamps from all over the world, I eventually concentrated on U.S. commemorative postage stamps. I was not a collector for any monetary reason, but like W. B. Yeats, I just loved the stories they told about the country. Commemorative stamps reinforced aspects of American history and geography and introduced me to people, events and places that were part of the American story.

Stamp collecting, more formally known as philately, has its own collection of terms and definitions. Here are some examples. Definitive or regular stamps were first issued in 1857 honoring both Benjamin Franklin, the first Postmaster General, and George Washington, the first U.S. President. These regular issues contained many national symbols—flags, the Statue of Liberty, presidents and many other famous Americans. In contrast, commemorative stamps were first issued in 1893 to celebrate the Columbian Exposition in Chicago and continued to celebrate famous

T. L. Bell (✉)
Department of Geography and Sustainability, University of Tennessee,
Knoxville, Tennessee, USA
e-mail: tombell@utk.edu

S. D. Brunn (ed.), *Transdisciplinary Research Across Knowledge Worlds*,
https://doi.org/10.1007/978-3-032-19491-6_13

Americans, noted landmarks and landscapes, and to recognize important events, organizations, international alliances, and environmental issues. Postage stamps have been described both as historical messengers (Brunn, 2011) and a form of "banal nationalism" (Billig, 1995).

In this chapter, I focus on four major changes that have taken place since I began collecting and suggest a fifth area of change that needs to be examined in future research. The four changes are: (1) the increase in bureaucratic oversight of stamp issuance; (2) changes in printing technologies; (3) the increase in volume, diversity, and inclusivity in stamp issuance; and (4) alterations in the mix of topical coverage in commemorative issues. The final part addresses the need for research on the impacts of digital alternatives such as email, social media and online payments as they relate to the future of the postage stamp.

Bureaucratic Changes Affecting the Issuance of Commemorative Stamps Since 1957

Coincidentally, the year I started my stamp collection was also the year that Postmaster General Arthur E. Summerfield (1953–61) created a Citizens' Stamp Advisory Committee (CSAC). It consisted of 15 experts from a variety of academic disciplines and fields of endeavor in the arts, sports, sciences, technology, and history. They meet quarterly in private sessions to select the 30 or so new commemorative postage stamps that will be issued annually. Today, there are approximately 30,000 annual requests from the general public to issue stamps honoring a plethora of organizations, landmarks, concepts, and distinguished people. There are rules and procedures that must be followed for any stamp issue to be considered; those are clearly stated in the Committee's guidelines. As of 2018, no individual, except a recently deceased U.S. President, can be honored until at least 3 years after their death. There is also a 3-year time lag between the time the idea for a stamp issue is submitted and its appearance as a stamp. A rejected proposal can be resubmitted for consideration, but only once. The most important criterion is that the proposed stamp will "primarily feature American or American-related subjects [or] if the subject had significant impact on American history, culture or environment." Another catch-all statement that is subject to interpretation suggests that there will be "a balance of stamp subjects that includes themes of widespread national appeal and significance" (Citizens' Stamp Advisory Committee, 2023).

A second major change to the operation of the post office took place in 1971 when the Postmaster General was no longer a member of the President's cabinet but rather the head of a branch of the Executive cabinet renamed the United States Postal Service (USPS). It was charged with a dual mission—to continue to improve the efficiency of the mail delivery system as a vital communication link both within the country and abroad, and to operate like a business that attempts to maximize its revenue stream. This second mission, I thought, would exacerbate the problem of stamp pandering—the issuance of colorful stamps on popular topics that would appeal to collectors both domestic and foreign but would not necessarily reflect the nationalistic patriotism of many stamps issued in the past. In the analysis discussed below, I discovered that stamps could serve two masters simultaneously—they can convey the history of our nation in a meaningful way and also provide colorful fodder for collectors of popular culture icons and topics as well.

Analyzing Changes in Commemorative Stamps Since 1957: A Thought Experiment

To analyze whether the patriotic/nationalistic story of America as told through commemorative stamps had been diluted and/or changed in any significant ways by changes in technology, selection criteria, and postal mission, I performed a thought experiment. What would a young person who began collecting U.S. commemorative postage stamps 50 years after I started and maintained his or her enthusiasm for the hobby take away from the commemorative stamps issued from 2007 to 2010? Would their story of America be significantly different from mine and, if so, in what way(s)?

I created two spreadsheets that I labelled "historic" and "contemporary" and recorded every U.S. commemorative stamp issued for the two periods: 1957–60 and 2007–10. Data were gathered on items such as: (1) a major theme (famous person, landmark, flora and fauna, international orientation); (2) the presence or absence of patriotic symbols (flags, maps, coats of arms); (3) whether the stamp was part of a designated series; and (4) whether it was commemorating an anniversary of a past event and, if so, which one. The data were derived from the *2017 Scott Specialized Catalogue of United States Stamps and Covers* (Houseman, 2016), the most authoritative source for assessing the value of U.S. postage stamps.

Initial Results of the Thought Experiment

Developing the spreadsheet for the historic period was no problem as the data in Table 1 reveal. Only 86 stamps representing 75 different thematic issues were issued in this 4-year historic period. But doing the same for all 411 stamps from the 109 thematic issues issued in the contemporary period proved a daunting and, quite frankly, useless task because of the multiplicity of stamps found on many of the souvenir sheets (i.e., mini sheets with many different stamps contained on a sheet) issued during the contemporary period (2007–10). Some of these souvenir sheets contained as many as 20 separate stamps. Examples of souvenir sheets in this chapter are Fig. 4a–d, 7a, 8, 9, 11, and 12. The multiplicity of stamp topics on a souvenir sheet is reflected by the differences between the number of topical **issues** of focus

Table 1 Historical stamp summary (1957–1960)

	1957		1958		1959		1960	
Category	Issue	Stamps	Issue	Stamps	Issue	Stamps	Issue	Stamps
Holidays								
Scientists/Diplomats			2	5	2	2	4	4
Artists			1	1				
Pop Culture/Toons								
Environment	1	1	1	1	1	1	2	2
International	2	2	4	7	4	5	10	14
Semi-Postal								
Landmarks			3	3	1	1	1	1
Commemoration	2	2	3	3	1	1	2	2
National Symbols	4	4	3	3	1	1	7	7
Sports								
Organizations	6	6			2	2	5	5
TOTALS	*15*	*15*	*17*	*23*	*12*	*13*	*31*	*35*

Table 2 Contemporary stamp summary (2007–2010)

	2007		2008		2009		2010	
Category	Issue	Stamps	Issue	Stamps	Issue	Stamps	Issue	Stamps
Holidays	5	13	5	13	7	16	4	13
Scientists/Diplomats			1	4	3	11	1	4
Artists	3	3	6	25	2	2	5	14
Pop Culture/Toons	6	45	6	21	4	27	4	11
Environment	5	28	3	12	2	12	2	11
International			1	1			2	2
Semi-Postal			1	1				
Landmarks	3	8	1	2	2	7	1	2
Commemoration	2	2	1	1	4	7		
National Symbols	5	7	2	64	2	5	2	2
Sports			1	1			2	2
Organizations	1	1					2	11
TOTALS	*30*	*107*	*28*	*145*	*26*	*87*	*25*	*72*

included in each of the two periods vis-à-vis the number of individual **stamps** that were produced to address each topical issue (see Tables 1, 2 and 3 for Historical, Contemporary, and Temporal Change by Stamp Category).

Thematic Shifts from the Historic to the Contemporary Period

More thematic issues and individual stamps were produced in the contemporary period. Not only had the quantity of stamps changed, but also the relative proportions of stamps assigned to one of 12 mutually exclusive and exhaustive categories

Table 3 Temporal change by stamp time period

Category	1957–1960 % of Total Stamps	2007–2010 % of Total Stamps
Holidays	0%	13%
Scientists/Diplomats	13%	5%
Artists	1%	11%
Pop Culture/Toons	0%	25%
Environment	6%	15%
International	33%	<1%
Semi-Postal	0%	<1%
Landmarks	6%	5%
Commemoration	9%	2%
National Symbols	17%	19%
Sports	0%	<1%
Organizations	15%	3%
TOTALS	*100%*	*100%*

changed in significant ways as well. Table 3 indicates the relative proportions of stamps that were produced in each of 12 thematic areas employed for this thought experiment. There were no holiday stamps produced in the historic period nor were there any popular culture/cartoon, semi-postal, or sports-related stamps produced. A semi-postal stamp is relatively new in the U.S. (the first one issued in 1998) but has been a long-time staple of many nations in Europe and elsewhere in which an amount of the stamp's purchase price goes to a designated charity and the remainder to defray the cost of postal services (the normal first-class postage rate).

One seemingly unexpected outcome of the analysis is the large proportion of the historic stamps devoted to international organizations and affairs (33 percent) compared to the meager proportion of stamps in the same category during the contemporary period (less than one percent). There were many more stamps issued to demonstrate that the United States was an important member of international groups and stamps that commemorated the anniversaries of international treaties as well, for example, the opening of the St. Lawrence Seaway in 1959 by issuing a stamp jointly with Canada. In the historic period of interest (1957–60), the post office honored 20 different international treaties and events with 28 individual stamps whereas in the contemporary period the USPS issued only three internationally oriented stamps. The most likely reason was the need to reinforce America's ascendence as a superpower in the Cold War era. And the issuance of stamps reflecting America's participation in international organizations during the historic period (1957–60) was a form of banal nationalism (Billig, 1995). Such stamp issues subtly announced the emergence of the U.S. as a player on the international stage, a position that needed no stamp promotion by the contemporary period (i.e., 2007–10). This shift to more nationalistic insularity of themes on commemorative stamps is not unusual. Brunn and Bell (2023) pointed out that the more affluent Western European nations that they examined (Netherlands,

Switzerland, Norway, and Italy) now also focus their attention more on domestic matters than international. This finding even applies to popular culture issues such as cartoon characters, popular movies, television, and music icons unique to the individual country.

By contrast, stamps with an environmental theme currently are twice as prevalent in the contemporary period compared to the historic period. Concerns about species extinction and global climatic change were barely on most people's radar in the 1950s. For example, Rachael Carson's *Silent Spring* on the dangers of the pesticide DDT was not published until 1962 (Carson, 1962). But by the early 2000s, environmental concerns were front-and-center as topics of concern and the production of stamps on these topics reflected the public interest.

The proportions of stamps honoring/commemorating personages in the arts, letters, and entertainment vis-à-vis science, diplomacy and politics also have switched. In the historic period, there were more than twice the proportion of stamps devoted to diplomats, political leaders, and those responsible for scientific advances compared to the contemporary period and vice versa for achievements in the arts, letters, and entertainment in the contemporary period in which more than double the number of such stamps compared to those honoring scientists, diplomats, and those in the political arena. Upon closer inspection, the reversals also reflect gender differences. The historic period honoring 27 diplomats, politicians and scientists was dominated by the Champions of Liberty series (issued from 1957–61) and all the persons in this category were male. Female representation improved in the contemporary period when Civil Rights advocates were included as approximately 30% of the 24 individuals assigned to this category were female. Although there were only five persons in the arts, letters and entertainment category who were honored on stamps of the historic period, three (60%) were female. Fifty-nine people representing the arts, letters, and entertainment category in the contemporary period were honored with stamp issues of which about 70% (42) were male.

The Impact of the USPS's Dual Mission on Stamps Reflecting U.S. Popular Culture

After the post office transitioned to the dual aims that guided the United States Postal Service (USPS) after 1971 (i.e., public service and commercial enterprise), stamps with popular themes and subjects, especially TV, music, movie stars and cartoon characters began to proliferate (Gallagher, 2016). Fully a quarter of all stamps in the contemporary period (2007–10) fit this description. The reasons for some of these shifting proportions are detailed below.

The Proliferation of Contemporary Holiday Stamps

There were no special holiday stamps printed in the historic period (1957–60). The first Christmas stamp, for example, was not printed until 1962. It was a simple candle and wreath design. In the contemporary period both religious and secular Christmas stamps are issued annually. Whether there was any objection to the decision to produce religiously oriented stamps remains unknown. Any discussions on the subject that might have transpired among members of the CSAC are not a matter of public record. The Christian religious stamp is usually based on a famous artistic rendition of either Madonna and child or a manger scene and there are several secular stamps issued both on sheets and in booklets. In 1996, the first Hanukkah stamp was issued followed by the first Kwanzaa stamp in 1997 and Eid in 2001. In the contemporary period (2007–10), 2008 was the first year the USPS first issued a stamp celebrating Chinese New Year. But holiday stamps are also now issued for Valentine's Day; the first Love stamps were issued in 1973). Annual wedding stamps started in 2006 and are issued in two denominations before June, the most popular month for wedding ceremonies. The higher denomination was likely issued to cover the extra cost of postage when RSVP envelopes are included.

Changes in the Degree of Subject Matter Specificity

Commemoration of state centennials and other state celebrations in 50-year intervals has become more "generic" as illustrated by Brunn (2018). Brunn and Bell (2021) highlighted the differences between the 1946 Iowa statehood centennial stamp festooned with symbols, flag, and map specific to the state and the state's 1996 sesquicentennial issue which simply depicted a Grant Wood painting of a generic farm scene he called "Young Corn" (Fig. 1). Though Wood was from Iowa, the scene could have been almost anywhere in the Midwest.

Brunn (2018) demonstrated this increasing trend toward the generic in statehood issues by comparing many state stamp issues over time. In the contemporary period that I examined, if the Oklahoma centennial stamp issued in 2009 did not have the name of the state and "Oh What a Beautiful Mornin'", a song from the musical "Oklahoma," written on a stamp, one would be hard pressed to identify the state that was being commemorated (Fig. 2). The stamp depicts a generic sunrise over a relatively flat riverine landscape that could be found in several states. That state birthday commemorations now appear more generic, (i.e., less likely to include people and events in the state's history of limited interest to non-residents than in years past), may relate to the commercial mission of the USPS. The newer state birthday stamps focus more on the physical environment than on historic battles and famous personages associated with the history of the state being commemorated than state birthday stamps issued previously. The use of colorful photogravure and lithographic printing processes compared to the more monochromatic engraved stamps of the

Fig. 1 (**a**) 1946 Iowa Centennial Stamp (Scott 942); (**b**) 1996 Iowa Sesquicentennial Stamp (Scott 3088)

Fig. 2 2007 Oklahoma Statehood Centennial (Scott 4121)

past has made this change possible and has, perhaps, widened the market for such stamps far beyond their appeal to residents of the state being commemorated.

The reverse, however, appears to be the case for other stamp subject areas. Environmental conservation stamps were generic and rather bland in the 1950s despite one of the first applications of more than two colors on commemorative stamps. This innovation was made possible by using the Giori press rather than the more common rotary press of the era. Shown in Fig. 3 are the four conservation stamps issued during the historic period (1957–60), each of which is printed in four colors.

In the 50-year interim after 1957, the environmental subject area became much more specific and colorful thanks to the photogravure printing process used to produce, for example, the series of 12 habitat-based souvenir sheets containing stamps representing the flora and fauna of the ecological habitat. They were issued annually from 1999 to 2010. Depicted in Fig. 4 are the four "Nature of America" souvenir sheets issued in the contemporary period I analyzed: Alpine Tundra (2007), Great Lakes Dune (2008), Kelp Forest (2009), and Hawaiian Rain Forest (2010), all vivid contrasts to the environmental themes depicted in the historic period.

Fig. 3 (**a**) 1957 Wildlife Conservation (Scott 1098); (**b**) 1958 Forest Conservation (Scott 1122); (**c**) 1959 Soil Conservation (Scott 1133); (**d**) Water Conservation (Scott 1150)

Fig. 4 (**a**) 2007 Alpine Tundra (Scott 4198 a-j); (**b**) 2008 Great Lakes Dune (Scott 4352 a-j); (**c**) 2009 Kelp Forest (Scott 4423 a-j); (**d**) 2010 Hawaiian Rain Forest (Scott 4474 a-j)

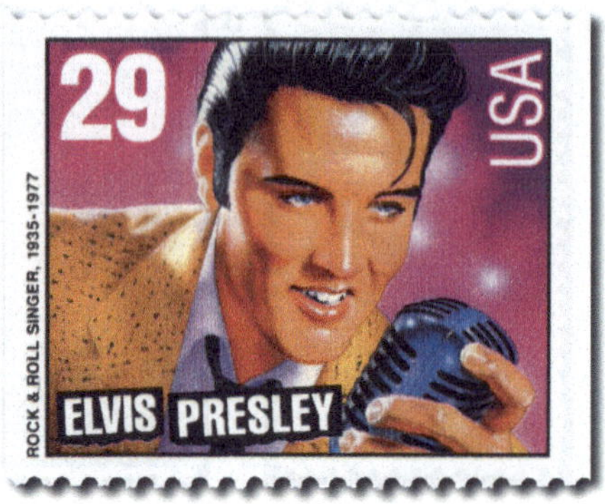

Fig. 5 The most popular USPS stamp issue was the 1993 Elvis Presley stamp (Scott 2721)

Each of the 12 "Nature of America" series souvenir sheets designed by Ethel Kessler contain ten different stamps cleverly hidden like a *Children's Activities* magazine picture puzzle. Each of the ten stamps in the souvenir sheet could be used as first-class postage. But most people buying such beautiful environmentally based souvenir sheets would be reticent to break them up and use them as postage. And that is exactly what the USPS had in mind—issuing stamps and souvenir sheets that are so collectable, such as adorable cartoon characters, famous actors, musicians and artists, that they would provide revenue for the USPS from their sales without imposing many costs, that is, there would be little or no need for processing and delivery hence almost pure profit for the agency. The estimated percentage of the most popular U.S. commemorative stamp ever issued (the Elvis Presley stamp of 1993 shown in Fig. 5) that was not used for postage is quite high—about 124 million of the 500 million that were printed (National Postal Museum, 2020).

Patriotism is Alive and Well in the Contemporary Period, but Internationalism Appears to Be on the Decline

In the historic period one set of issues, the Champions of Liberty series, seems to have violated the Citizens' Stamp Advisory Committee (CSAC) rules that U.S. stamps should depict only Americans or foreign persons who were instrumental in aiding and abetting the American cause such as the Frenchman Marquis de Lafayette. Of the ten freedom fighters selected for inclusion in this series issued from 1957 to 1961, only Lajos Kossuth, the Hungarian freedom fighter who moved to the U.S. after Austria overthrew his government in 1851 and who addressed a joint session of Congress, had close U.S. ties. So why were the other nine so honored? The answer appears to be that even the CSAC and the Postmaster General must comply with the President's wishes. A recent email from a private stamp company wishing to sell collectors complete sets of the Champions of Liberty series states:

> As America was embroiled in the Cold War, the Eisenhower administration saw these stamps as a way to combat Soviet totalitarianism by honoring men who fought for freedom in their homelands. Each of the people honored in the Champions of Liberty series struggled courageously to protect their homeland from tyranny and further the cause of freedom.
>
> After [Ramón] Magsaysay's single commemorative stamp was issued, the rest of the series had two stamps printed for each subject—one in a single color, and one with three colors. The pairs were also of different denominations. The single color (sic) stamps covered the domestic rate, while the multi-color stamps paid the international rate, to carry the message of freedom around the world (Mystic Stamp Company, 2023).

As mentioned above, there were many more stamps issued to demonstrate that the United States was an important member of international organizations and to commemorate the anniversaries of international treaties in the historic period of interest (1957–60), than in the contemporary period. As Brunn and Bell (2023) have pointed out, this increase in nationalistic insularity is not an unusual phenomenon. Many advanced countries concentrate on issuing stamps that commemorate their own important historical events, citizens and culture first and foremost.

A concern that was allayed comparing the results of the two periods is that despite the many pandered issues authorized by the USPS in the contemporary period, the story of American achievement, heroism and patriotism is still presented effectively and colorfully. Also, because of the increase in the number of commemorative stamps issued by the USPS made possible by improvements in printing technology, the contemporary young person collecting U.S. commemorative stamps would be informed to a much greater extent about historical events, landmarks and famous people than his or her historical counterpart. I feel that the two major contributors to this rich historical content are: (1) the creation of several ongoing thematic series that are more wide-ranging and inclusive of previously underappreciated and diverse groups and genres; and (2) the nostalgia for the many beautifully engraved stamps of the past that have been reformulated and reprinted to highlight philatelic events. In the contemporary period (2007–10) stamps from 12 different series were issued. If that period were extended to the present, even more nationalistic series would be added. These contemporary series include the Black Heritage Series, Literary Arts Series, American Treasures Series, Legends of Hollywood Series, American Scientists, American Journalists, American Landmarks, Flags of Our Nation, Civil Rights Pioneers, Ecological Habitat Series, Distinguished Sailors, and Negro Leagues Baseball.

Stamp Gimmicks Used to Boost Sales are Nothing New While New Printing Technologies Greatly Expand the Possibilities

Following the contemporary period, even more daring printing gimmicks have been employed to boost sales of stamp products. To celebrate the 50th anniversary of the moon landing in 2019, for example, a stamp depicting a reflection in astronaut Buzz

Fig. 6 2019 Mirrored Image in Aldrin's Space Helmet (Scott 5399)

Aldrin's space helmet was printed on chrome paper to emphasize and enhance the mirror image being reflected (USPS, 2019) (Fig. 6).

At about the same time (2018–19), the USPS introduced two souvenir sheets printed on lenticular paper that is biconvex like the lens of the human eye. The Art of Magic (2018) souvenir sheet shows a magician's top hat with a rabbit appearing and disappearing depending on the viewing angle (Fig. 7a). The same lenticular printing process was used in 2019 to create an animation-like effect on a souvenir sheet depicting different life stages of a T-Rex dinosaur including a menacing adult that appears to move (Fig. 7b).

There Is Nostalgic Appeal for Stamps of the Past as Well

Cognizant that postal history issues are popular, the USPS has produced several reproductions of early engraved stamps giving collectors and the general public the opportunity to view anew the best of U.S. postal history. Two popular issues shown here celebrate the centennial of the date of the original issue. The first is a 1998 souvenir sheet of the nine stamps of various denominations designed to emulate the 1898 issues commonly referred to as the Trans-Mississippi West stamps that experts argue are the finest examples of stamp engraving ever produced (Fig. 8). The purchaser of this souvenir sheet might be reticent to remove and use an individual stamp for postage even though in theory they could.

The second homage to the past is a 2013 issue celebrating the centennial of perhaps the most famous U.S. postal error ever discovered—the inverted Jenny stamp

Fig. 7 (**a**) 2018 Lenticular Disappearing Rabbit Images in The Art of Magic (Scott 5306); (**b**) 2019 Lenticular Image of a Menacing T Rex (Scott 5411)

(Fig. 9). Originally issued as a 24-cent airmail stamp, one 50-stamp sheet of the bi-colored stamp was sold in which the airplane itself was mistakenly printed upside down before other such sheets were discovered and destroyed. The USPS was savvy enough to issue the souvenir sheet honoring the stamp error's centennial with each of the six stamps on the sheet carrying a $2.00 denomination. Once again, few buyers of the $12.00 sheet would wish to use a single stamp from the souvenir sheet as postage. But the USPS went a step further to gain interest in the relatively expensive souvenir sheet. The agency purposefully created its own rare error. As stated in the 2017 Scott Catalog (Houseman, 2016), the souvenir sheet "along with a piece of gray cardboard backing, was inside a sealed plastic wrap. One hundred panes were produced that contain the airplane right-side up. These panes were included in the same envelope and outerwrap (sic) and were distributed somewhat randomly." The value of the souvenir sheet with the plane right-side up to the lucky individuals who received it was $75,000 in 2017.

The volume of stamps issued in the contemporary period increased tremendously because of newer, faster and more colorful printing methods replacing the tediously produced engraved stamps of earlier eras. The last engraved stamp issued by the USPS was issued in 2005 to honor the centennial of the birth of the Swedish-born legendary film actress Greta Garbo (1905–1990). The engraving was based on a 1934 black and white photograph of the movie star's face and was issued jointly with Sweden.

Fig. 8 1998 Trans-Mississippi West Reissue (Scott 3209 a–i)

Also, a great tribute has also been paid to the appearance of such early engraved stamps by emulating them using more modern printing techniques. Not only has the USPS reproduced stamps issues of the early period of America's postal history, but they have also honored famous Americans in our early history with new stamps meant to appear like those that might have been engraved more than a century ago. The Lewis and Clark bicentennial issues of 2004 illustrate this engraved "look" very well (Fig. 10).

Pandering Issues that Stretch or Violate the CSAC's Own Rules

If the CSAC rules were tweaked a bit during the historic period (1957–60) for Cold War geopolitical purposes, they were stretched even more during the contemporary period (2007–10) examined and continue to be stretched to the present. But importantly, as the title of his chapter implies, the patriotic and pandered stamps have never been separated into completely different tracks to form their own parallel

Fig. 9 2013 Centennial Miniature Sheet of Most Famous U.S. Postal Error—Upside Down Jenny (Scott 4806)

Fig. 10 2004 Lewis and Clark Issues (Scott 3854–56)

universes. There have always been, and will continue to be, intersections between the historically patriotic topics and the contemporary popular and more inclusive and diverse topics.

One might point out the introduction of cartoon characters as an example of pandering for sales that began in 1995 with the 20-stamp souvenir sheet entitled "Comic Strip Classics" depicting images of U.S. newspaper comic strips from the "Yellow Kid" to "Brenda Starr." One could also argue that the chronology of the American newspaper comic strip itself is an important aspect of popular culture history. Pandering and the history of popular culture icons thus intersect and coexist.

In 1997, the first stamps featuring cartoon characters much beloved by Americans (and everyone else for that matter) featured Warner Bros. characters (Bugs Bunny, Daffy Duck) depicted with variations on the theme of a mailbox. This was followed in 2004 with stamps showing Disney characters that were issued almost annually depicting a different theme (friendship, celebration, romance, imagination). In 2006, DC Comics (e.g., Superman, Batman) joined the cartoon explosion and in 2007 these superheroes were joined by members of the Marvel Universe (Spiderman, The Hulk, Iron Man). The use of popular cartoon characters to sell many stamps and generate revenue for the USPS is what Brunn and Bell (2023) and others (e.g., Bell & Senn, 2023; Silvester, 2023) have called stamp "pandering."

Brunn and Bell (2023) extended the notion of pandering by coining the term "geopandering," largely based on an article with the provocative title "Why Is Elvis on Burkina Faso Postage Stamps? Cross-Country Evidence on the Commercialization of State Sovereignty." (Slemrod, 2008). In this article, Slemrod uses stamps containing Disney characters, Princess Diana, and/or Elvis Presley to define operationally a stamp pandering nation—issuing many stamps not based on the geographic origin of the stamp's subject matter. Such geographically out-of-place pandering is aided and abetted by private companies that produce and distribute stamps for many poorer countries that cannot afford to print and distribute their own stamps. Forty-two such stamp pandering countries were so defined (Slemrod, 2008). While most were small and poor countries in the Caribbean, Africa and Oceania, the United States itself was also so defined!

Some of the stamps issued by the USPS border on geopandering. One might wonder why, for example, John Lennon, who was born in the UK, was one of the musicians to be honored in the Musical Icons series in 2018 (Fig. 11).

The answer is that his music influenced many American musicians and thus he should be included. The promotional literature issued by the USPS in 2018 when the stamp was announced reads:

> *The world was influenced by John Lennon's music and his commitment to the ideals of peace and unity. We continue to be inspired by his memory. The Music Icons: John Lennon … stamp features a black-and-white photo of Lennon taken by noted rock 'n' roll photographer Bob Gruen in August 1974.* The original photograph has been treated in gradations of color… Lennon's signature appears at the top of the stamps. "USA," the peace symbol, and the Forever denomination appear along the bottom. The stamp pane is designed to resemble a vintage 45 rpm record sleeve. One side of the pane includes the stamps and brief text about Lennon's legacy, with the image of a sliver of a record seeming to peek out the top of the sleeve. (USPS, 2018)

So, I will give the USPS a mulligan for the inclusion of a non-U.S. citizen of such enduring importance to American music.

A more recent stamp issue of 2021 called the "Message Monsters" (Fig. 12), in my opinion, irreparably breaks the rules set forth by the CSAC and could also be considered an example of geopandering, that is, pandering in which geography is out-of-place.

Fig. 11 2018 John Lennon Issue Music Icons Series (Scott 5315)

As stated in the USPS website promoting the souvenir sheet's launch:

> The U.S. Postal Service will celebrate Message Monsters with the most playful, customizable Forever stamp design ever. The four monster illustrations on this pane of 20 stamps invite interactivity with dozens of self-adhesive accessories on the selvage. The monster-ific accoutrements include cartoony voice balloons and thought bubbles with exclamations and salutations, hats and crowns, hearts, stars, crazy daisies and other fun flair. Art director Antonio Alcalá designed the pane with original artwork by Elise Gravel, author and illustrator of popular children's books. (United States Postal Service, 2021)

What the USPS promotional literature fails to mention is that the illustrations are from the famous Canadian author and illustrator Elise Gravel. So, while the intention of the stamp issue to encourage young people to become more interested in stamps and mailing letters is admirable, especially from a stamp collector's perspective, I fail to see how that intent is particularly or peculiarly American.

Fig. 12 2021 Message Monsters Sheet with Non-postal Selvage Stickers (Scott 5636–39)

The Future of the Commemorative Postage Stamp: The Need for More Research

While I do not wish to end the chapter on a sour note, I cannot ignore the "elephant in the room." That elephant is the uncertain future of the postage stamp itself. In 2020, Iceland made the decision to stop printing new postage stamps altogether and Finland's postal authority suggested that it was also considering doing the same. Iceland will use the stamps that it has on hand and, if demand necessitates, will reprint older stamps (McCarty, 2020). As with many advanced and highly digitized countries in Europe and North America, the sales of postage stamps have plummeted, increasing the costs for postal delivery. Denmark is another country that will stop the delivery of letters at the end of 2025. The number of letters handled by the

Danish postal service (Post Nord) has declined from 1.4 billion in 2000 to 110 million in 2024 (Murray & Kirby, 2025). The increase in digital communication alternatives (e.g., email, social media, online bill payment) has significantly reduced the volume of letters being mailed.

What are the alternatives to the postage stamp? Some countries (e.g., United Kingdom, Germany) now include QR codes as selvage on their postage stamps. The QR codes could eventually replace stamps altogether with advances in automated scanning equipment. Likewise, postal credit accounts in which people are assigned a unique identifier that would replace the postage stamp are being discussed. Whatever the future holds for the commemorative postage stamp and the hobby of stamp collecting, these topics are topic worthy of further investigation.

Summary and Conclusions

A comparison of U.S. commemorative postage stamps issued between 1957 and 1960 (the historic period when I was an avid stamp collector) with stamps issued 50 years later (between 2007 and 2010) as the contemporary period were compared to determine how the story of America as told through its stamp issuance has changed. And, of course, the philatelic American story itself has changed...mostly for the better in my opinion. Contemporary stamps are more numerous, colorful, playful, and exotic but also many are every bit as patriotic as those issued 50 years earlier if a tad less international (i.e., more nationalistically insular). As America emerged in the post-war period as the acknowledged superpower of the free world, the relationships it forged with other nations in that sphere were highlighted in stamps honoring bilateral and multinational alliances including the Universal Postal Union (UPU), vital to international communications. As an alternative to totalitarian regimes in the emerging Cold War era, two issues during the historic period were especially important—the Champions of Liberty series honoring freedom fighters confronting authoritarian regimes throughout the world over two centuries and the American Credo issue of 1960–61 featuring six patriotic quotations from famous American statesmen and founding fathers (e.g., Washington, Jefferson, Franklin).

There were more stamps issued in the historic period commemorating the anniversaries of commercial entities and trade organizations, such as centennials of the American Institute of Architects and the Petroleum Industry, but the contemporary period was much more likely to honor contributions in both the sciences (especially aerospace) and fine arts than 50 years earlier. Many of these tributes were part of a designated series, such as Black Heritage, American Scientists, Distinguished American Diplomats, and those so honored were a more diverse and inclusive group than those on stamps in the historic period.

The contemporary period is also filled with cartoons and popular culture icons because of the necessity of the USPS to increase revenue to the extent possible by the sale of stamps and stamp-related products (e.g., clothing items featuring stamps, framed stamp posters honoring new releases). These ancillary stamp products are readily available through USPS philatelic services. During the COVID-19 pandemic, even several stamp-related jigsaw puzzles were marketed by the USPS because of the rising interest in stay-at-home family-based activities. But the pandered stamps and stamp products produced by the USPS are not separate like a parallel universe from issues commemorating important people, places and events in the evolving American story. Both types of issues intersect almost as a cross-pollination of the past and present. Many of the popular culture issues including Black Cinema, Jazz, Blues, Tejano Music, and Hip Hop, honor philatelically the overlooked contribution of persons of color of both earlier and contemporary periods.

While most contemporary stamps are unlikely to appreciate much in monetary value, they continue to tell an interesting, informative, and colorful story of both current events as well as doing great honor to the past and to the iconic symbols of our nation. That patriotic homages to the past coexist with modern popular culture on contemporary stamps is fine with me as it provides a broadening opportunity for the present generation to learn about both the present and the country's past with cartoons and amusements thrown in for good measure. But the future of postage stamps in the United States and elsewhere in light of advances in digital technologies such as social media, email, and the like is a problematic matter for a separate research project.

References

Bell, T. L., & Senn, J. (2023). Local political upheaval and global cultural appropriation: The Postage Stamps of Grenada, 1974-1988. In S. D. Brunn (Ed.), *Stamps, nationalism and political transition* (pp. 312–330). Routledge.

Billig, M. (1995). *Banal nationalism*. Sage.

Brunn, S. D. (2011). Stamps as messengers of political transition. *Geographical Review, 101*(1), 19–36.

Brunn, S. D. (2018). The state branding of U.S. postage stamps for state commemorative years. In U. Ermann & K.-J. Hermanik (Eds.), *Branding the nation, the place and the product* (pp. 29–52). Routledge.

Brunn, S. D., & Bell, T. L. (2021). *Geopandering: Forthcoming article*. Ubique, Newsletter of the American Geographical Society.

Brunn, S. D., & Bell, T. L. (2023). Geopandering. *Geographical Review, 113*(3), 433–448.

Carson, R. (1962). *Silent Spring*. Houghton Mifflin.

Citizens' Stamp Advisory Committee. (2023). *Who We Are*. What We Do. United States Postal Service. Accessed at https://about.usps.com/who/csac/

Gallagher, W. (2016). *How the post office created America: A history*. Penguin Books.

Houseman, D. (2016). *2017 Scott specialized catalogue of United States Stamps and Covers*. Amos Media.

McCarty, D. (2020). *A new dead country: Iceland to stop issuing stamps*. Linn's Stamp News. Accessed at https://www.linn.com/news/us-postal-history/a-new-dead-country-iceland-to-stop-issuing-stamps-may-8-2020

Murray, A., & Kirby, P.. (2025, March 7). *Denmark Postal Service to stop delivering letters*. BBC. Accessed at https://www.bbc.com/news/ckg8;llq283

Mystic Stamp Company. (2023). Champions of liberty series. Accessed at https://info.mysticstamp.com/this-day-in-history-august-31-1957.

National Postal Museum. (2020). The Elvis Stamp: America elects a king. Accessed at postalmuseum.si.edu/exhibition/art-of-the-stamp-the-artwork-stamps-with-a-story/the-elvis-stamp-america-elects-a-king.

Silvester, W. (2023). Cartooning Islands for fun and profit: The proliferation of Disney stamps. In S. D. Brunn (Ed.), *Stamps, nationalism and political transition* (pp. 331–349). Routledge.

Slemrod, J. H. (2008). Why is Elvis on Burkina Faso Postage Stamps? Cross-country evidence on the commercialization of state sovereignty. *Journal of Empirical Legal Studies, 5*(4), 683–712.

United States Postal Service. (2018). Imagine…John Lennon Forever. Accessed at https://about.usps.com/news/national-releases/2018/pr18_069.htm

United States Postal Service. (2019). U.S. Postal Service issuing 2019: First moon landing forever stamps. Accessed at https://about.usps.com/newsroom/national-releases/2019/0610ma-first-moon-landing-stamp.htm.

United States Postal Service. (2021). Message Monsters Ready to Bring a Smile to Your Mail. Accessed at https://about.usps.com/newsroom/national-releases/2021/0811ma-new-message-monsters-forever-stamps-feature-a-customizable-design.

Yeats, W. B. (1926). *Coinage Bill speech*. Irish Free State Senate.

Montserrat's Parallel Universes: Mapping Science, Policy and Society During a Volcanic Emergency

Omari Graham, Arlette Saint Ville, and Gabrielle Thongs

> My home is in Kinsale. My home is over there, all my things [are] over there. I can't go over there. I live in–in the shelter now. I have to buy coals to cook my food because my stove is at home in Kinsale. We don't know how to live. We don't want to go England; I don't want to go. I have nobody over there. I want to stay with someone but I would rather stay in Montserrat. If I die. I die.—(Resident) Anelle Kabbee, Montserrat (Personal Communication, August 1997)

> In the context of risk, there is considerable scope for arts-based approaches that use interactive and visual methods to communicate risk and uncertainty and that emphasize that communication also involves listening.—(Researcher) Amy Donovan (2019)

Introduction

Disaster Risk Governance Gaps

Disaster risk governance involves a web of multi-level actors across formal and informal institutions that interact within complex networks to formulate and implement polices to manage disaster risk. One way disaster risk governance systems interact with the public is by communicating changes in both absolute and relative perceptions of risk levels across society (Wilkinson, 2015). One component involves the functioning of the disaster pipeline from warning to action.

The complex yet fascinating nature of volcanoes and volcanic eruptions has within the last century or so given birth to a very dynamic scientific discipline broadly referred to as physical volcanology. Modern volcanology is considered by

O. Graham
The University of the West Indies Seismic Research Centre,
St. Augustine, Trinidad and Tobago
e-mail: omari.graham@uwi.edu

A. S. Ville (✉) · G. Thongs
Department of Geography, Faculty of Food and Agriculture, The University of the West Indies, St. Augustine, Trinidad and Tobago
e-mail: arlette.saintville@uwi.edu; gabrielle.thongs@uwi.edu

S. D. Brunn (ed.), *Transdisciplinary Research Across Knowledge Worlds*,
https://doi.org/10.1007/978-3-032-19491-6_14

most practitioners as an innately multidisciplinary field. Typically, scientists and researchers in the field have backgrounds in geology, chemistry, physics and other physical sciences. These scientists tend to focus on improving scientific understanding of the processes before, during and after eruptions. The diversity of volcanic environments, eruptive styles and processes remains a rich source of curiosity for researchers and has the potential to fuel inquiry for another hundred years. However, the propensity of humans to exploit natural environments coupled with the allure of the volcanic landscape has given rise to another type of volcano scientist—or at the very least, a scientist—with parallel interests. When humans settle near volcanoes, the question of risk that must be assessed, managed and governed arises. Social volcanologists, as they are called, hail from the physical as well as the social sciences and recognize that volcanic processes also impact people.

Continued human settlement near active volcanoes has given rise to the parallel universes of scientists who study volcanoes in pursuit of the esoteric and those whose quest for knowledge is driven by a need to protect society from the impacts of eruptions. In the literature, the interaction between these at times diverging pursuits has been described as the hazard-risk (Bretton et al., 2018) or science-management (Jolly & Cronin, 2014) boundary or interface depending on the author's leaning. For those who believe science must be protected from undue influence, the place where these worlds meet remains a boundary. For others who recognise and appreciate the value of contextualized science to risk management, disaster risk governance and communication, the intersection operates at an invaluable interface. There is good reason to keep the "science" and "management" related to volcanoes in separate spheres. When public/government responses to natural hazard events are based on scientific advice, the separation between scientific roles and those responsible for enacting mitigation action is thought to shield scientists from direct liability for casualties and other losses. The imprisonment of six scientists accused of providing inaccurate information prior to the deadly L'Aquila earthquake in 2009 (Aspinall, 2011; Bretton et al., 2015; Scolobig, 2015), is often cited by those who advocate for maintaining this separation. Conversely though, in some circumstances, a closer relationship between "science" and "management" can be beneficial.

The Caribbean region, a collection of mostly independent small island developing states is the second most disaster-prone region in the world (CDEMA, 2014). One of the defining characteristics of small island developing states is their relative lack of human resources, particularly in highly specialized fields. The Caribbean region is no different when it comes to scientific capacity and this is particularly true for volcanologists. Although the Lesser Antilles chain of islands is the product of active volcanism and there are 21 potentially active volcanoes in the region, the number of professionally trained volcanologists working there remains relatively low. It is mainly for this reason that scientists at the University of the West Indies Seismic Research Centre (UWI-SRC), the scientific agency that monitors volcanoes on the English-speaking islands in the region, has adopted a policy of bridging the gap between science and management. Doing anything else would only serve to increase the vulnerability of these islands to eventual volcanic catastrophe.

When the Soufriere Hills Volcano reawakened in 1995, scientists from the UWI Seismic Research Unit were the first to arrive on the island to support authorities there. Established in 1953 by British colonial authorities to respond to volcanic emergencies in the Caribbean (Latchman et al., 2012), the unit by the time of the Montserrat crisis had already supported the government of neighbouring St. Vincent and the Grenadines through the 1979 eruption of the La Soufriére volcano in that country. In Montserrat, the UWI-SRC has managed the Montserrat Volcano Observatory since 2008. During this time UWI-SRC scientists based at the observatory and experts they collaborate with have pursued their scientific work mindful that the information they provide has the potential to impact lives in a very real sense.

Volcanic hazard and the risk pipeline connected to the Soufrière Hills Volcano in Montserrat has been assessed in a consistent and quantitative manner for over 14 years (1997–2011) by multiple levels of specialists. These efforts include the Scientific Advisory Committee (SAC) (and predecessors) in collaboration with the Montserrat Volcano Observatory (Wadge & Aspinall, 2014). The committee (SAC, 2025) is commissioned by the Overseas Territories and Polar Directorate of the UK's Foreign, Commonwealth and Development Office. The SAC is responsible for assessing the scientific integrity of the work being carried out by the Montserrat Volcano Observatory. Members meet twice a year, once on Montserrat and once by teleconference.

The Disaster Risk Governance Gap: Communicating Changing Volcanic Risk

In a World Bank report on early warning systems associated with volcanic hazards, Watson (2018) acknowledges that over time monitoring of volcanoes to inform forecasting and early warning of volcanic activity has been significantly enhanced by technological advances. Tools mentioned range from increased satellite images on deformations and remotely sensed imaging on thermal and gas emissions in the data collection phase, to use of open-sourced databases and modelling during data processing and analysis. These aids have improved our understanding of volcanic processes, supported by enhanced data accessibility in regional and global databases. The primary way that scientists communicate these risks with local populations is with volcanic alert levels represented by a simple color, letter or number code that report the assessed level of volcanic activity. Shared primarily through national government authorities, volcanic activity levels represent the assessed volcanic risk level based on key elements of the volcano's status (e.g., level of unrest, time scale, eruption likelihood, indicative phenomenon, recommended action). They are tailored to the characteristics of the volcano and locally based hazard assessment maps that often link volcanic activity levels to exclusion zones (Watson, 2018). While volcanic risk levels are informed by increasingly sophisticated tools from monitoring networks, direct observations, satellite imagery, census data and

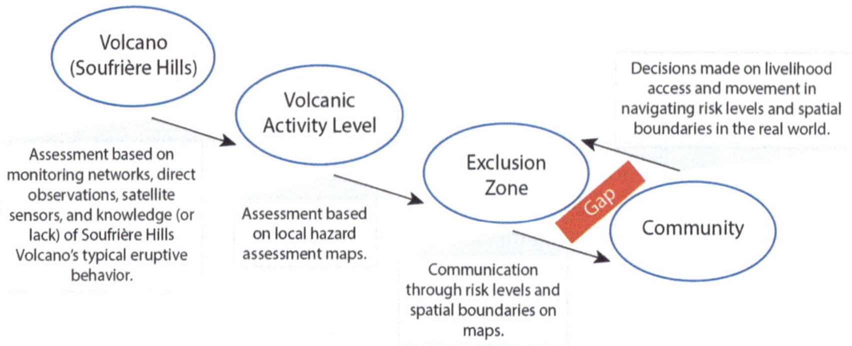

Fig. 1 Flow chart illustrating the decision-making process related to the Soufrière Hills Volcano. It begins with "Volcano (Soufrière Hills)" leading to "Volcanic Activity Level," assessed through monitoring networks and technical observations. This connects to "Exclusion Zone," determined by local hazard maps. A gap is identified between the "Exclusion Zone" and "Community," where decisions on access and movement are made. Communication with the public occurs through changing risk levels and spatial boundaries on maps

knowledge (or lack) of a volcano's typical eruptive behaviour, there are gaps related to the local population's engagement with volcanic activity levels and exclusion zones. This photo essay seeks to fill in this gap for the Soufrière Hills Volcano in Montserrat (Fig. 1).

Exclusion zones represent a spatial display of decisions promulgated by disaster officials to communicate volcanic risk levels and limit movement in designated high-risk areas while confirming safe zones. Often, these lines drawn on the ground or on a map may not be well-appreciated by politicians, visitors and residents. At worse, the risk perception may not be shared by residents in threatened areas. This means that defining of exclusion zones requires more than lines drawn in the sand but ongoing dialogue and negotiation across the science–policy divide to communicate effectively to residents and impacted communities (see Fig. 1).

Method

The unpredictable Soufriere Hills Volcano located on the British Overseas Territory of Montserrat provides a background to understand how initial, small and incremental, as well as rapid escalation of volcanic activity, has given rise to successive forced evacuations as guided by scientists and a multiplicity of policy changes initiated to communicate risk via the exclusion zones.

We use photographs from the UWI Seismic Research Centre and the Montserrat Volcano Observatory from 1997 to 2024 to illustrate events along the disaster pipeline from warning to action. We describe the sequential steps taken by hazard specialists, policymakers, and disaster communicators to mitigate the impact. The images provide key snapshots of the changes to the volcano and shifting exclusion zone around Montserrat's volcano. We use secondary documents to inform us about the volcano's history and physical changes and mapping of human geographical changes as well as details on how government policies were developed and how they influenced individual decisions to move across the island.

All maps have been redrawn from the original Montserrat Volcano Observatory maps. Existing maps and satellite imagery were sourced from established records and academic literature to show spatial changes over time in line with the photographic themes and timelines.

Since the Soufrière Hills Volcano's geomorphological activity caused significant changes to the island's coastline, particularly around 2010, digital elevation models were created to represent these changes spatially. Using editing and Geographic Information System (GIS) tools and techniques, pre-2010 and post-2010 digital elevation models were created to address these issues. Subsequently, further GIS tools were used to digitise the previous exclusion zones on georeferenced maps. They were then attributed to the respective volcanic risk time period. The digitised zones were then overlaid on the digital elevation models and shapefiles of local (past and present) communities. The spatial layers were overlaid depending on the appropriate time period to assess how the zones shifted and to spatially represent how it affected the population's exposure and patterns of land use.

In this research, GIS was initially used to spatially compare historical static maps. To facilitate a comparison of static hardcopy maps across the volcanic timeline, each historical map was georeferenced to Montserrat's local coordinate system. This effort enhanced the integration of several geographic datasets and optimised a comparison of the exclusion zones as well as the movement of inhabitants responding to the fluctuating risk levels. Moreover, by spatially displaying the geographic datasets, it was also possible to observe the distinct alterations to the coastline that gradually occurred. In multiple areas across the island, the shoreline was expanded by the continuous deposit of pyroclastic flow material. A spatial analysis of these historical maps and current satellite imagery provided a powerful visual representation of this geomorphological activity that may not be readily striking in person.

Results and Discussion

Early Responses to the New Volcanic Hazard

Since inception, the Soufrière Hills Volcano exclusion zones and their respective maps have been shaped by an ongoing collaborative effort led by the Montserrat Volcano Observatory and enriched by scholars and analysts from the scientific community. With each iteration, they have been strengthened by their unique insights, particularly by geologists. Hicks and Few (2015) explain that early risk maps were developed by late 1995 and simplified in September 1997 (final revision in April

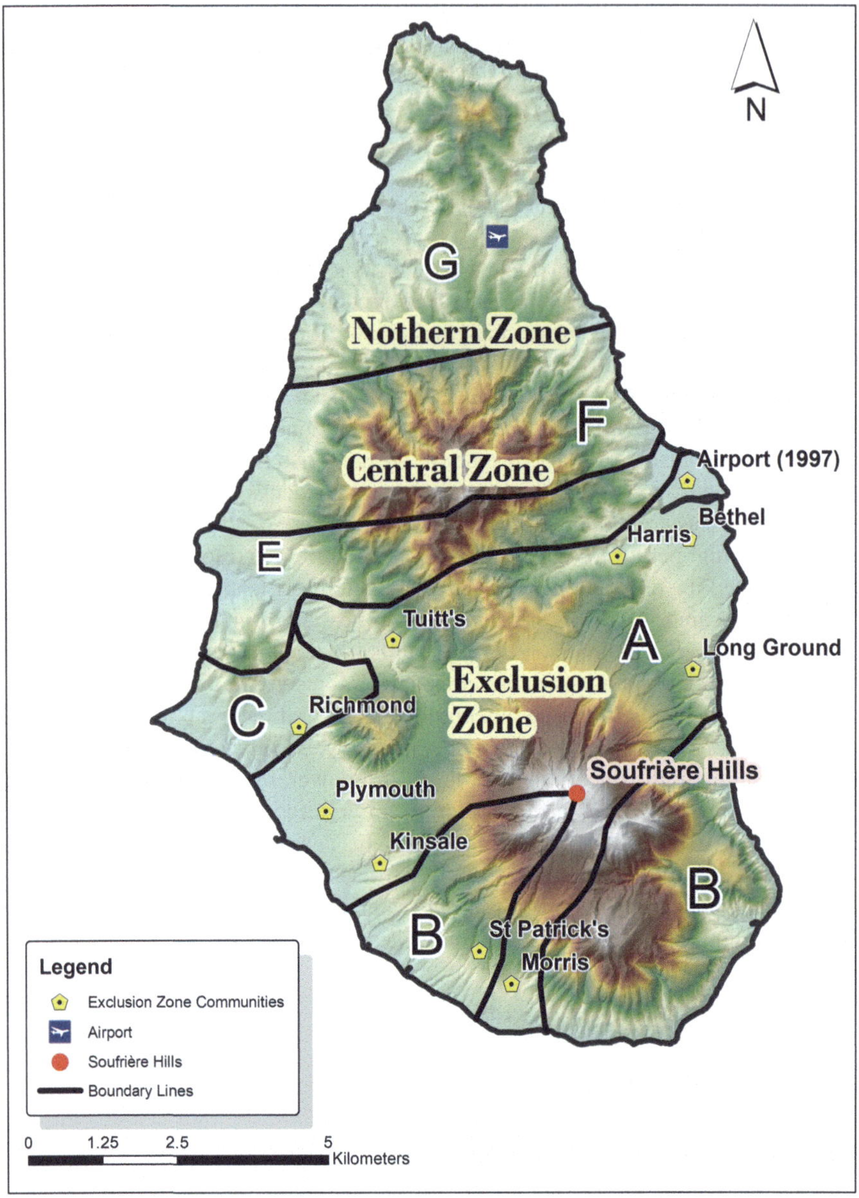

Fig. 2 Map of the geographical region divided into zones: Northern, Central and Exclusion. These reflect the earliest 1997 exclusion zone boundaries. (Adapted from Hicks & Few, 2015)

1999). Figure 2 illustrates the early responses in communicating the increased volcanic risk levels in 1997 and provides a brief snapshot of an early iteration of an exclusion zone. At this time, the island-dwellers had limited experience with this hazard. Zone A, the highest risk exclusion zone, covers the area of closest proximity to Soufrière Hills Volcano. Zone A is located in the southern half of the island and represents the area assessed most at risk to the active volcano and the corresponding pyroclastic flows. Over time, these exclusion zone boundaries shifted as the lava dome grew. In this iteration of the exclusion zone, B and C were located to the south and west of Zone A. Zones B and C were transitional areas that served as much needed buffer zones that were always in a high state of alert because of their relatively close proximity to the highest risk exclusion zone, Zone A.

After the 1996 heightened risk levels, residents in Zone A from the city of Plymouth and other communities in the south (Zone B), were internally displaced and re-settled to the North of the island. In other areas movement was controlled and in some of the buffer areas, movement was prohibited. These variable exclusion zones were established to control access to areas close to the volcano according to the assessed level of volcanic activity (see Fig. 2).

There were two sets of evacuations in 1999, from the capital city of Plymouth following the initial volcanic activity. After short periods, residents resettled in homes. Following the movement and return, there was no formal exclusion zone. In 1999 (Fig. 3) there was a significant adjustment to the exclusion zones based on changes in the assessed risk levels. At this time the entire southern half of the island was reassessed and categorized as the exclusion zone. This major adjustment led to the evacuation of many more communities which were previously considered to be of moderate risk based on their proximity to the volcano. With the new classification, communities in the South of the island were now regarded as the highest risk locations with the highest possibility of impact. Some of these newly assessed high-risk communities were Morris, St Patrick's, Kinsale, Harris, Long Ground, and Montserrat's capital, Plymouth.

These and other major risk management decisions are listed in the timeline found in the Discussion section. The governance arrangements and relationships shaping these decisions and collective responses to volcanic risk provide a lesson on how these risks were communicated. Some of the results have been studied elsewhere and are discussed below.

Kokelaar (2002, p. 01), on summarising the nature of the Soufriere Hills Volcano eruption timeline, noted some challenges that informed the period 1997–1999:

> The slow progress and long duration of the volcanic escalation, coupled with the small size of the island and the vulnerability of homes, key installations and infrastructure, resulted in a style of emergency management that was dominantly reactive .... Based on frequent hazard assessments, a series of risk management zone maps was issued by administrative authorities to control access as the eruption escalated. These were used in conjunction with an alert-level system.

The end result from the perspective of the community was unnecessary loss and difficulty for residents. Kokelaar (2002:01) states:

> The unpreparedness of the Montserrat authorities and the responsible UK government departments resulted in hardship, ill feeling and at times acrimony as the situation deteriorated and need for aid mounted. Losses and stress could have been less if an existing hazards assessment had registered with appropriate authorities before the eruption.

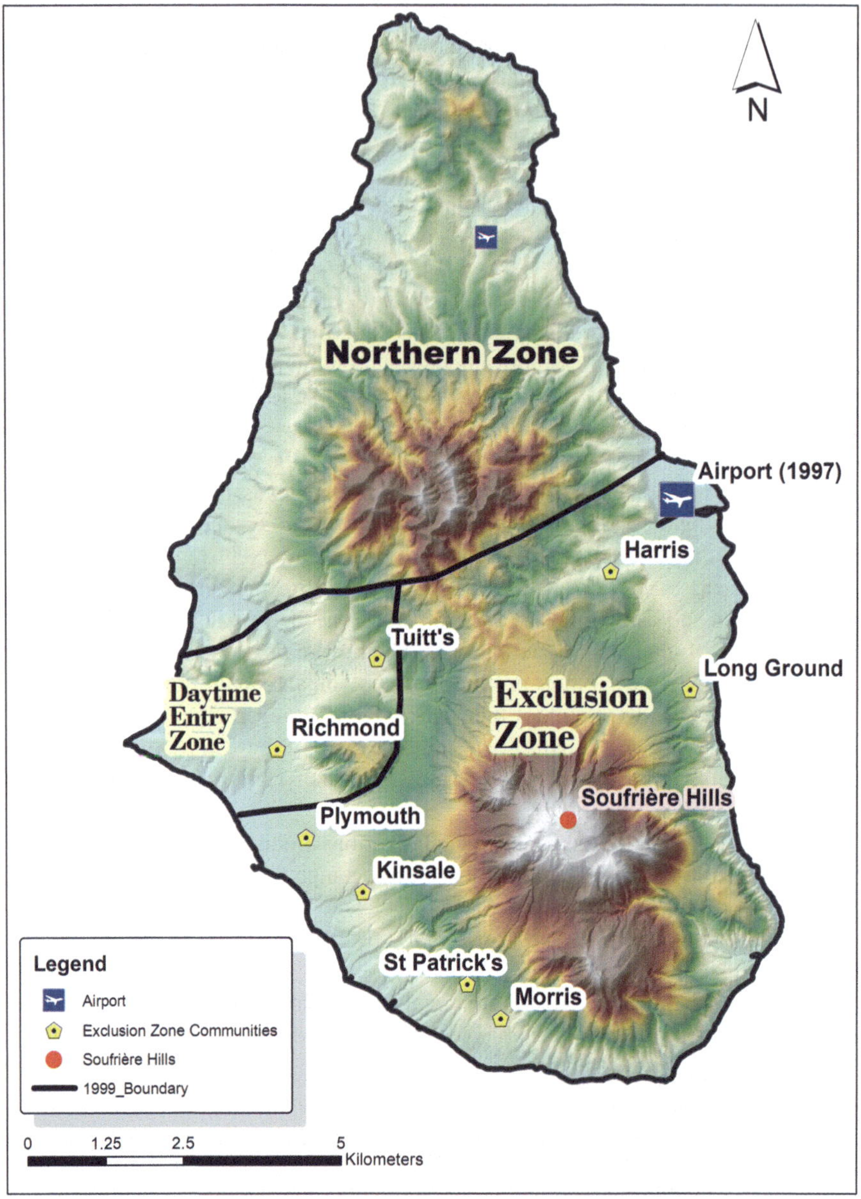

Fig. 3 1999 map of Montserrat showing major towns and cities, critical facilities and the Soufriere Hills Volcano. (Adapted from Hicks & Few, 2015)

In September 1996, the seven 'microzoned' hazard zones (A-G) served to communicate the risk status of each zone. These zones were modified based on the various alert levels issued by the government of Montserrat as recommended by Montserrat Volcano Observatory volcanologists. This report communicated that in the G and F Zones covering the northern parts of the island people were safe when the volcanic risk level was at its highest. In contrast, many of the other areas located in the south that were closest to the volcano (Zones A & B) could not be accessed, or if accessed, were only suitable for short visits allowable during the times when the volcanic dome was in a stable growth phase (Hicks & Few, 2015).

Earlier in 1996 certain areas were deemed accessible to the public when the volcano was characterised by the scientists at the lowest alert level. Understandably, over time there was a need to revise and review the alert scheme and construct maps to communicate to the public the ongoing changes to the hazard risk. These multi-layered spatial representations of risk that were communicated through the disaster pipeline from volcanologists to policymakers and the community as lines on a map gave rise to "misunderstanding between the public, civil authorities and the scientists, and occasionally disregard to follow official warnings" (Aspinall et al., 2002:219). Despite overall efforts that limited movement to reduce loss of life and harm, on 25 June 1997, disregard for danger from the eruption proved deadly and resulted from primarily smallholder farmers returning to farms at the base of the volcano. Nineteen persons were killed when pyroclastic flows destroyed settlements and infrastructure (including the airport in the East) from Trants to Dyers (Loughlin et al., 2002). In August 1997 the exclusion zone was expanded with three communities (Salem, Old Towne and Frith) being evacuated. At this time, there was a move to a clearer and more simplified messaging with three broad zones. Zone D was merged into Zone C following the events of 25 June.

The nature of the volcanic crisis involved first an internal displacement period with inadequate shelter that was overcrowded and involved the forced separation of many families with women and children sleeping in shelters and mature male members elsewhere, including in vehicles. In 1996, approximately 6000 of the 7000 Montserratians who left the island relocated to the UK under the British governments' Voluntary Evacuation Scheme as part of the volcano relief package (Shotte, 2007).

These post 1995 migrants moved to North and East London. In their study on psychosocial effects and coping with the disaster by Montserratians who relocated to the UK as a result of the Montserrat volcanic disaster, Barnes (2003) interviewed 60 participants and organized focus group discussions with evacuees in Birmingham and London from 1997 to 1998. Interviewees noted that early migrants in London were supported by black evangelical churches. These churches engaged the relocated community and organized various social gatherings such as parties, dinners, and games during daytimes and evenings to maintain and strengthen social networks. The reason given for their relocation to cities like Preston, Leicester, London and Birmingham was that these areas were locations where post-war Montserratian migrants had previously settled (Barnes, 2003).

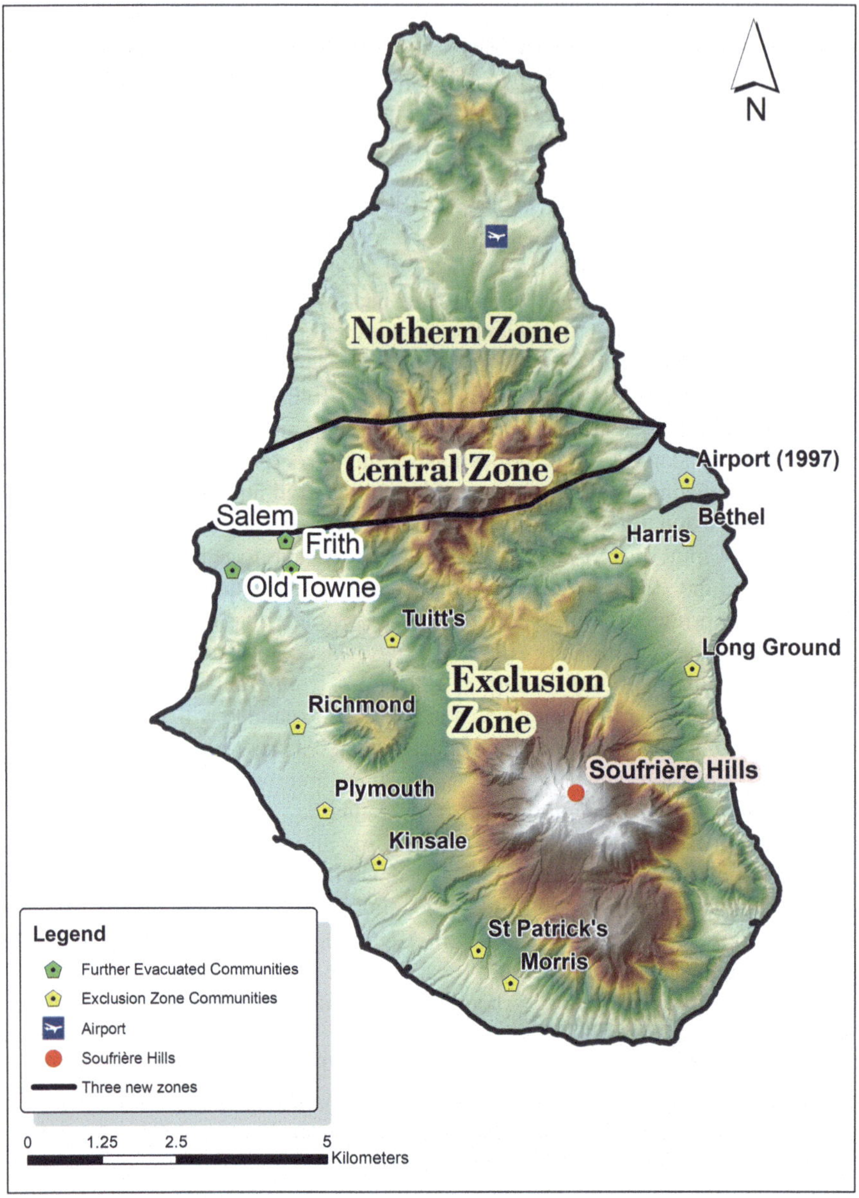

Fig. 4 After the 1997 deaths, the zoning map showing return to simplicity of messaging with three zones: Exclusion, Central and Northern (Adapted from Montserrat Volcano Observatory, 1997)

In Fig. 4, the three broad zones were identified and simply communicated the potential hazard within the exclusion zone covering the Soufriere Hills Volcano, Plymouth and its surroundings. The central zone previously C & E (became known as the Daytime Entry Zone) and the Northern Zone which covered the areas that were considered safe (G, F, & E). The three-tiered exclusion zones remained in place until 2008. After 2008 (final revision in 2011), a new hazard level system was implemented accompanied by a new hazard map. The new map involved a further subdivision of the southern part of the island (primarily the Exclusion and Central Zones). At this point, two-thirds of Montserrat was divided into five zones (A, B, C, F & V) and also included two Maritime Exclusion Zones (W & E). Access to some of these zones, particularly A and B, depended on the alert level (0–5 with 5 being the highest alert level) and rendering zones A-D inaccessible.

By 2011 the volcanic hazard zones became more intricate and were communicated through colored access restricted zones lettered A, B, C, D, E, F and V (Figs. 5 and 6).

Results

Here we use a series of photos to depict physical and social impacts of the Soufrière Hills Volcano eruption.

Physical Changes Due to Volcanic Activity

Figure 7 shows the destruction of an evacuated village within the exclusion zone early in the eruption. Pyroclastic flows are among the most dangerous hazards associated with volcanoes that erupt explosively. These fast-moving mixtures of hot gas, ash and rock easily reach speeds over 100 km/h and consume everything in their path for up to tens of kilometers. The nature and scale of this volcanic hazard confine risk management strategies and recommended actions to evacuations and the strict enforcement of exclusion zones in settlements located around the volcano. Despite the existence of these exclusion zones during the height of the volcanic crisis in Montserrat, pyroclastic flows claimed the lives of 19 people who had returned to unsafe portions of the island to tend to their farms. Regions on the east of the island (see Fig. 7) including communities like Trants and Dyers which were destroyed by flows in July 1997. Like its Caribbean neighbour Mt. Pelée in Martinique where *nuée ardente* (pyroclastic flows) were first described and documented in modern times, the Soufriere Hills Volcano played an important role in scientists' understanding of this complex phenomenon. Videos of pyroclastic flows or Pyroclastic Density Currents (PDCs), as they are now more commonly called, travelling over water were first captured during the Soufriere Hills eruption on Montserrat.

In Fig. 8, a mix of gas and ash rises into the air over Soufrière Hills Volcano. Volcanologists usually class volcanic hazards as either proximal or distal depending on their reach. Proximal hazards generally refer to those with impacts experienced not more than 20 km from the vent. These hazards include ballistics or projectiles

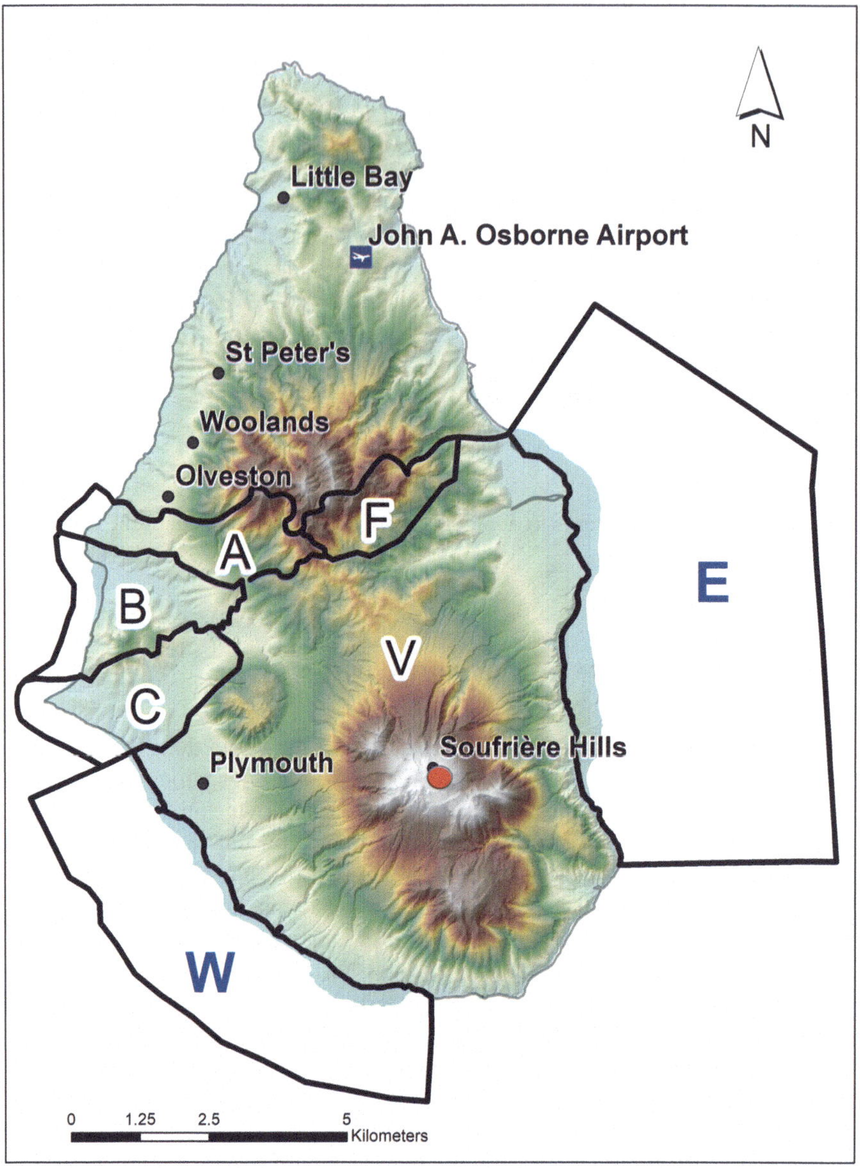

Fig. 5 The zones displayed in this map were implemented in August 2008 (http://www.mvo.ms/pub/Hazard_Level_System/). At this point, Montserrat was microzoned into six areas (A, B, C, F, T & V) and included the two maritime exclusion zones (W & E). (Adapted from Sword-Daniels et al., 2014)

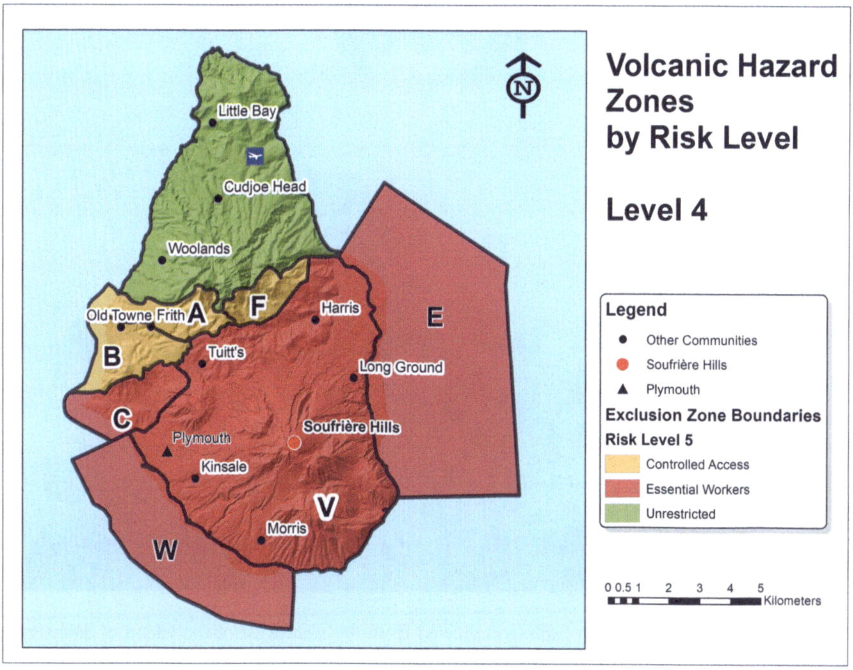

Fig. 6 Map showing the most updated exclusion zones in Monserrat by risk level, where 5 (red) represents the highest risk, orange the medium risk and green the lowest risk levels

Fig. 7 A pyroclastic flow engulfs an evacuated village in Montserrat during the early stages of the Soufriere Hills eruption circa 1996 (Photo Credit: Montserrat Volcano Observatory/UWI-SRC photo archive)

Fig. 8 Soufrière Hills Volcano explosion viewed from the southwest of the island of Monserrat (Photo Credit: Montserrat Volcano Observatory/UWI-SRC photo archive)

thrown from the vent during explosions as well as most pyroclastic flows, lateral blasts and toxic gas emissions. Distal hazards are those that impact locations beyond 20 km from the volcano and invariably may have an equal and often greater disruptive effect on human activity.

The human settlements around Soufrière Hills Volcano pictured in Fig. 8, were initially evacuated and eventually abandoned due to the risk posed to those who would have previously occupied these areas. Historically, the flanks of volcanoes have always attracted human settlement. Fertile soil, the by-product of long forgotten eruptions, offers the promise of plentiful and profitable agriculture in areas close to volcanoes. On small islands, where flat lands suitable for large-scale physical development are often scarce, the gently sloping land created by the type of volcanic activity found at Soufrière Hills Volcano often invites settlers/developers to make significant investments in property and livelihood. These investments, as occurred in the South of Montserrat, however, eventually became exposed when Soufrière Hills Volcano entered the destructive phase in its ongoing cycle of creation and loss.

Even as they threaten human activity, geologic events, particularly volcanic eruptions, are a poignant reminder that the natural processes that have shaped the current environment we inhabit continue unhindered. Risk management approaches that consider volcanic hazards must, therefore, make provision for the eventual continuation of geologic processes that may significantly impact the natural and built

Fig. 9 New land created on the east coast of Montserrat from pyroclastic material moving over the flanks of Soufrière Hills Volcano. (Photo Credit: Montserrat Volcano Observatory/UWI-SRC photo archive)

environment. Figure 9 depicts the formation of new land along Montserrat's east coast from material ejected from the currently active volcano. The new land is built on the material deposited by previous eruptions from several volcanic centres. Montserrat is dominated by three major mountain ranges from the oldest in the north to the youngest in the south: the Silver Hills in the north, the Centre Hills and the Soufriere Hills in the south. Together these three ranges reveal a story of land shaped by alternating episodes of volcanic cataclysm and calm. During the periods of calm, humans either through ignorance or disregard for nature, placed themselves, their livelihoods, and cultural landscapes on a collision course with processes that have occurred for millennia. While the human species remains on earth, the challenge of interacting with a planet that continually reshapes itself by seemingly violent means will continue to engage those required to coexist with the same processes that make life possible. Modern scientific tools give contemporary society the ability to monitor and prepare for these episodes of destruction and renewal. These tools should ideally not only be used during crises to provide early warning prior to hazard impacts but also to guide long-term land use to minimize loss during periods of natural reconfiguration as seen in Fig. 9.

Human Impact, Social Adjustment and a New Way of Life

Figures 10 and 11 display the now evacuated primate capital city of Plymouth. Once the official capital of Montserrat, Plymouth has come to be known informally as the "Pompei of the Caribbean" after the devastation it suffered during the Soufrière Hills Volcano eruption. As the economic and cultural centre of a small island community, Plymouth housed many essential services and served as the administrative hub for the island's government. When the Soufrière Hills Volcano reawakened to life, the picturesque town with its colonial architecture stood directly in the path of the worst the volcano could muster. Montserratian institutions like Angelo's Supermarket suffered a fate few residents could see coming but was always in the cards. After being evacuated early on, Plymouth and its environs began to be impacted first by ash then by more and more destructive phenomena. Churches, banks, hotels, schools and cinemas in the once lively town have all saw their last parishioners, customers, visitors, students and patrons. After 30 years of volcanic bombardment what remains of Plymouth are the rooftops of once stately buildings entombed in layers of ash, rock and mud. While names like "The Attic Restaurant" survive in new locations further north, the landmarks of the old town are no more. The eruption forced Montserrat's social and cultural events to find new homes and

Fig. 10 War memorial in Plymouth in the process of being buried under ash and other volcanic debris. The memorial is now completely covered (Photo Credit: Montserrat Volcano Observatory/ UWI-SRC photo archive)

Fig. 11 A Montserrat Volcano Observatory technician walks past Angelo's Supermarket in the now evacuated Plymouth town centre (Photo Credit: Henry Odbert)

moved the seat of power and prominence from its perennial place to the hilly north of the island.

Even as Montserratians processed the loss of shared public spaces, places and memories, they were doing the same for their own homes and communities. In Fig. 12 the remains of the kitchen area stand in burnt ruins showing what remains after residents were evacuated. Having left their homes for what many were hoping would be a brief evacuation, the reality of never returning to their island village or street would have been difficult for many to accept. There is one anecdotal account of a man who because of his job had access to the exclusion zone during the heights of the eruption. On his visits to the restricted area, he would sit where his house once stood and weep without restraint. While officials can bravely try to quantify the value of the properties lost to the eruption, the emotional cost of those losing their sense of place can never truly be counted. When risk managers require people to leave their homes, they are doing much more than asking people to take mitigation action. In post-colonial societies of the Caribbean where ownership can often be reserved for certain portions of the populace, there remain deep emotional and spiritual ties in efforts to claim some piece of the soil as their own. Familial identities, cultural traditions and personal milestones, all tied to geographical locations, have been destroyed by Soufrière Hills Volcano and played host to the often mundane, but personally fundamental activities of daily life. Communications asking people to leave these foundations that have supported generations of lives must therefore reflect the gravity of the potential loss people may experience.

Fig. 12 The scorched remains of a home showing a kitchen area that was destroyed by a superheated pyroclastic flow. Flow temperatures can reach up to 900 °C incinerating everything in their path. (Photo credit: Henry Odbert)

Areas once off limits to reduce the risk to residents are now accessible again but with some restrictions. Current risk levels as assessed have resulted in reduced risk levels with recommended action by authorities that permit resettlement and commercial activities in areas closer to the volcano. These changes in the Exclusion Zone are not without their challenges. Although some areas may not now be directly threatened by the volcano, they are still impacted by enduring fallout from the eruption. Residences south of the Belham River on the western side of the island are constantly under threat of being cut off (Fig. 13).

Lahars or mudflows can cause significant disruption when they occur; typically, during the tropical rainy season. Ash deposited during previous eruptive episodes is remobilised during rainstorms turning river valleys into impassible tolls. Figure 14 illustrates the signage that serves to remind residents of the secondary hazard of mudflows and latent vulnerability of their community. People that have chosen not to abandon their homes beyond the Belham valley must always be prepared to be marooned for a few days or even weeks when the inevitable happens (Fig. 15). This has now become a way of life for those unwilling to surrender their piece of paradise. Living and working in the shadow of an active volcano may be a personal decision, but there remains a duty of care for those responsible for ensuring public safety. How these two at times of diverging interests are managed is the challenge faced by authorities managing this now 30-year eruption.

Fig. 13 Urban cultural landscape in southern Montserrat destroyed by the erupting Soufrière Hills Volcano marked by ash covered slopes. Rainfall and overland flow have visibly eroded the ash deposits, creating tiny channels and rills. Soufrière Hills Volcano is partly visible as a large overbearing shadow in the background (Photo credit: Henry Odbert)

Authorities on the Eastern Caribbean Island of Montserrat have chosen to manage volcanic risk using a 5-level volcanic activity level system and risk management zoning that divides the island into 7 zones. Access to these zones is granted or restricted based on the level of activity at the Soufriere Hills Volcano and the threat posed to residents. Communicating the limits, locations and hazard status of these zones is a critical component of the government's risk management strategy. Figure 15 shows the street sign for entry in Zone B, a community in an area of controlled access depending on the volcanic activity level. Residents must constantly be kept aware of which zone they live in, work in, and the precautions that apply to these locations at any time. In volcanic environments where things can change quickly, the tools that communicate these changes should be able to keep pace with the dynamic nature of volcanic hazard and risk. Digital tools like GIS that enable spatial information to be collected, analysed and shared in quick time should logically take a leading role in the communication of hazard and risk along the disaster risk management pipeline.

Fig. 14 Mudflow warning sign at a Belham River crossing cautions would be travelers. The Montserrat Volcano Observatory can be seen keeping watch in the background (Photo credit: Omari Graham)

Discussion

Importance of Resilience Building and the Role of Technology in Communication Messaging

Montserrat, the "Emerald Isle" of the Caribbean, bears the mental, social and landscape scars from Soufrière Hills Volcano's 1997 activity that covered a relatively short period. Resilience, flexibility and science–policy messaging evolved quickly over time (Fig. 16). The sophistication of the disaster governance architecture guided by volcanologists to assess volcanic risk levels, develop volcanic activity levels, and message these in a meaningful way to guide livelihood changes, human evacuations and eventual resettlement needs to be appreciated. That architecture did so with relatively low mortality over the 15-year period covering the most intense volcanic activity also needs further study. Figure 16 helps us to better understand navigating of the volcanic risk levels, volcanic activity zone changes and human responses. The need to interpret maps, believe the "science" and respond appropriately harkens back to a time preceding science skepticism and denial. The recent work on the Soufriere in Saint Vincent and the Grenadines suggests more than a

Fig. 15 A street sign on the road to Ilses Bay at the Hazard Zone B boundary warning of the entry along the road into Zone B(Photo credit: Omari Graham)

"name" connection between the "Soufrieres." Research that noted the successful evacuation and high responsiveness from the neighboring island La Soufriere suggests that the work, messaging and communication on pending hazards by the University of the West Indies Seismic Centre has incorporated many of the lessons learned from Soufrière Hills Volcano (Graham et al., 2022).

Return Migration and Rebuilding: Not Politics as Usual

Efforts from as early as 2010 proposed a charter for returning and overseas nationals involved a suite of concessions to actively incentivize nationals to return home to Montserrat (GoM, 2010). These efforts are ongoing. In the 4 October 2024, elections, 33 candidates were available at the polls from the four main political parties. Across the spectrum varying degrees of each party promoted return migration and its benefits. The People's Democratic Movement (PDM, 2024) party launched its 46-page manifesto on 16 October 2024 which described housing and lands programmes. Not to be outdone, the Movement for Change and Prosperity (MCAP, 2024) launched its 32-page manifesto, the Positive Progression for People (PPP,

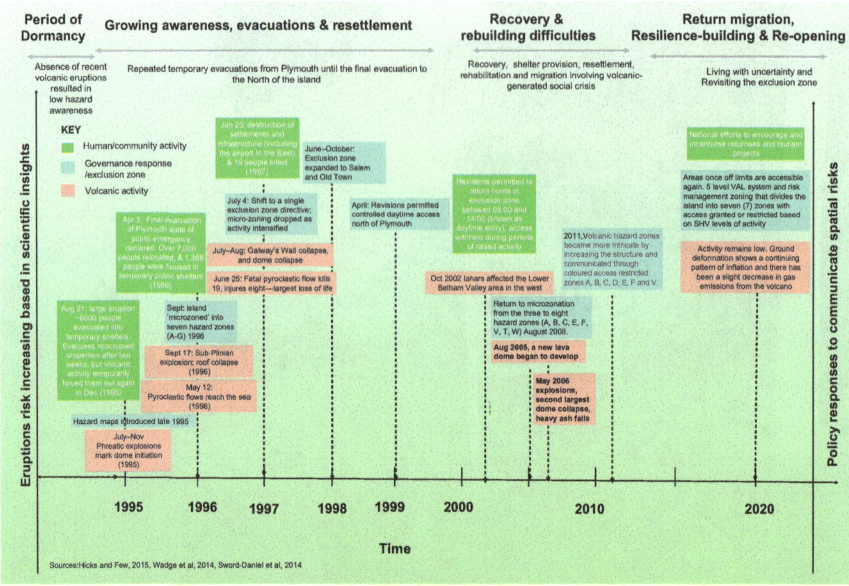

Fig. 16 Timeline flow chart illustrating volcanic activity and responses on Montserrat from 1995 to 2020. The chart is divided into four phases: "Period of Dormancy," "Growing Awareness, Evacuations & Resettlement," "Recovery & Rebuilding Difficulties," and "Return Migration, Resilience-building & Re-opening." Key events include eruptions, evacuations, and policy changes. Color-coded boxes represent human/community activity, governance response/exclusion zones, and volcanic activity. Sources: Hicks and Few (2015), Wadge and Aspinall (2014), Sword-Daniel et al. (2014)

2024) party its 10-page manifesto and the United Alliance (2024) a 12-page manifesto. The latter was victorious.

Improvements in housing stock were featured as one of the common propositions. Additionally, there were varied projects, activities and incentives to construct homes, attract and incentivize returning Montserratians. While political parties showed some innovation that involved Soufrière Hills Volcano there were some unexpected projects. Some projects proposed turning ash into cash, more conventional volcano tours, one promoted using the buried city of "Pompei" Plymouth as a tourist attraction and even re-establishing hotels in the shadow of the volcano. Returnees lured with shiny manifestos and incentivized by tangible benefits might need national public service announcements to help manage expectations of the uncertainty and changing volcanic activity levels related to sharing a small island with an active Soufrière Hills Volcano. Returnees will need to be informed of and coached toward living with acceptable risk and the psychological tools needed to accept volcanic activity levels and exclusion zones as scientifically flexible borders that are part of the new cultural and emotional landscape of the 102 sq. km island.

Geospatial Approaches to Enhance Risk Visualization

One of our key foci involved spatially displaying the evolution of volcanic activity levels and exclusion zones in Montserrat. These zones shifted and adapted in response to the dynamic volcanic risk on the island. As a result, they mirrored the varied intensities and spatial dimensions of volcanic risk to the inhabitants and focused on the existing scholarly literature of the impacts of the destruction (Hicks & Few, 2015; Monteil et al., 2020; Montserrat Volcano Observatory, 1997, 2023; Sword-Daniels et al., 2014; Wilkinson, 2015). The review revealed that GIS played a critical role in not only developing the foundational maps in the mid-1990s as eruptions occurred in Montserrat, but also in the ongoing contemporary risk and mapping efforts today. While exclusion zones in Montserrat were historically displayed using static hard copy maps, with efforts to attract returnees back home and enhance communication on risk, static maps are limited. It is expected that at times these maps will prove challenging for residents, returnees and visitors to understand their spatial location in reference to real time risk levels. Use of geospatial tools to create a dynamic map based on an individual's location in relation to the present risk level in Montserrat could be useful to its residents and visitors.

Using a spatial dataset, a dynamic map was built and uploaded in Google Earth (Fig. 17). It allows locals and visitors to view their live location relative to the current volcanic risk level in Montserrat. To ensure the map was easily accessible, two links were created. One takes the user to the browser version of the risk map (https://www.google.com/maps/d/embed?mid=1xzJ386Wn47dcKiO-N61ew_upCejx2Dw&ehbc=2E312F) while the other allows the user to directly open the Google maps application (https://goo.gl/maps/yeEFgAArsFTTW6hRA?g_st=ac).

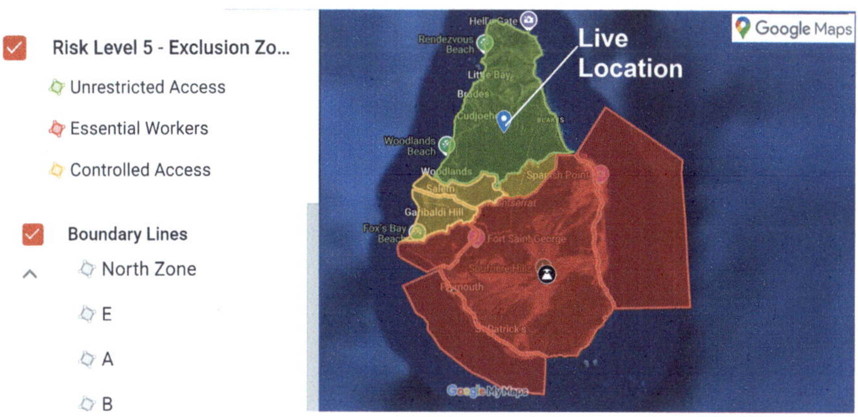

Fig. 17 This spatial data-driven application uses locals and visitors' actual location to show their position relative to Montserrat's volcanic risk levels and exclusion zones. The screenshot shows the live location (blue icon) of a resident by clearly displaying their location in an unrestricted area (Google Maps, 2025)

By analyzing live locations vs. the island's risk level, an individual's ability to grasp the locational-based danger drastically improves. These tools highlight how emerging technology can support early warning (Landwehr et al., 2016) and point to the need for a broader understanding of leveraging interactive platforms and social media in developing visual infographics for improved disaster communication and management. This application was designed and built to enhance individuals' perception of risk and, by extension, improve their self-protective behaviour. Critical hazard information is now accessible, easy to understand and surpasses traditional static representations.

Conclusion

This chapter examines the parallel universe across the science–policy interface that gives rise to changing exclusion zones as used by disaster officials to communicate volcanic risk levels, limit movement based on changing volcanic activity levels and identify scientifically supported safe zones areas in Montserrat. It shows that the lines drawn in the ground and on the map may not have initially been well appreciated by policymakers, visitors and residents, but after the loss of 19 lives simplifying these lines gave rise to further dialogue and negotiation between scientists, authorities and the resident communities. The data involved a timeline of volcanic eruptions, human movement and Exclusion Zone changes and constructing maps of the evolving exclusion zone footprint using remotely sensed satellite imagery, photos and secondary data to show human-society responses. The findings showed both dynamism and increasing complexity in the communication of risk as rapid and sustained rises in the level of volcanic risk that led to frequent changes in the identified zones placed within and without the footprint of the Exclusion Zone. Main themes taken from the chapter include resilience building important and role of technology in communication messaging, return migration and rebuilding, not politics as usual, and innovative geospatial approaches to enhance risk visualization. Future research should focus on the value of island "smallness" and proximity of science and management actors to effective disaster risk communication. The lessons gained from these approaches will likely be applicable to and welcome in other regions of small island developing states facing similar circumstances.

References

Aspinall, W. (2011). Check your legal position before advising others. *Nature, 477*(7364), 251–251. https://doi.org/10.1038/477251a

Aspinall, W. P., Voight, B., Baxter, P. J., Miller, A. D., & Lynch, L. L. (2002). Effective decision-making and risk communication in the Montserrat volcanic crisis. In T. H. Druitt & B. P. Kokelaar (Eds.), *The eruption of Soufrière Hills Volcano, Montserrat, from 1995 to 1999* (Geological Society, London, Memoirs, 21) (pp. 217–254). Geological Society of London.

Barnes, V. C. (2003). *Shepherding in ashy times: Pastoral care with Montserratian volcano evacuees in two Pentecostal churches in the UK*. The University of the West Indies. Retrieved from https://global.uwi.edu/sites/default/files/bnccde/montserrat/conference/papers/barnes.html [Vernie Bar…e with …]

Bretton, R. J., Gottsmann, J., Aspinall, W. P., & Christie, R. (2015). Implications of legal scrutiny processes (Including the L'aquila trial and other recent court cases) for future volcanic risk governance. *Journal of Applied Volcanology, 4*(1), 18. https://doi.org/10.1186/s13617-015-0034-x

Bretton, R. J., Gottsmann, J., & Christie, R. (2018). Hazard communication by volcanologists: Part 2 - Quality standards for volcanic hazard assessments. *Journal of Applied Volcanology, 7*(1), 10. https://doi.org/10.1186/s13617-018-0079-8

CDEMA. (2014). *Regional comprehensive disaster management (CDM) strategy and results framework 2014–2024*. Caribbean Disaster Emergency Management Agency.

Donovan, A. (2019). Critical volcanology? Thinking holistically about risk and uncertainty. *Bulletin of Volcanology, 81*(4), 20.

Google. (2025). Satellite image of the Soufrière Hills Volcano, Montserrat; 16.741993 N, −62.208443 W. Image data © 2024 Maxar Technologies, CNES/Airbus [Satellite image]. *Google Maps*. Retrieved August 10, 2025, from https://www.google.com/maps/d/embed?mid=1xzJ386Wn47dcKiO-N61ew_upCejx2Dw&ehbc=2E312F

Government of Montserrat. (2010). Handbook & guide for returning nationals and nationals in the diaspora. https://www.gov.ms/wp-content/uploads/2022/10/Montserrat-Diaspora-Handbook-Guide-for-Returning-Nationals-and-Nationals-in-the-Diaspora.pdf.

Graham, O., Edwards, S., & Robertson, R. (2022). Managing stakeholder relationships for improved situation awareness during volcanic emergencies: An Eastern Caribbean case study. *International Journal of Disaster Risk Reduction, 67*, 102656. https://doi.org/10.1016/j.ijdrr.2021.102656

Hicks, A., & Few, R. (2015). Trajectories of social vulnerability during the Soufrière Hills volcanic crisis. *Journal of Applied Volcanology, 4*(1), 10.

Jolly, A. D., & Cronin, S. J. (2014). From eruption to end-user; bridging the science–management interface during the 2012 Te Maari Eruption, Tongariro Volcano, New Zealand. *Journal of Volcanology and Geothermal Research, 286*, 183. https://doi.org/10.1016/j.jvolgeores.2014.08.025

Kokelaar, B. P. (2002). Setting, chronology and consequences of the eruption of Soufrière Hills Volcano, Montserrat (1995–1999). *The Geological Society of London, 21*. https://doi.org/10.1144/GSL.MEM.2002.021

Landwehr, P. M., Wei, W., Kowalchuck, M., & Carley, K. M. (2016). Using tweets to support disaster planning, warning and response. *Safety Science, 90*, 33–47.

Latchman, J. L., Lynch, L., & Edwards, S. (2012). 60 years of monitoring volcanic and earthquake activity in the English-speaking Eastern Caribbean. *Caribbean Geography, 17*(1&2).

Loughlin, S. C., F. V. Michael, A. D. Miller, G. E. Norton, K. C. Rowley, R. S. J. Sparks, & S. R. Young (2002). The Montserrat Volcano Observatory: Its evolution, organisation, role and activities.

Monteil, C., Barclay, J., & Hicks, A. (2020). Remembering, forgetting, and absenting disasters in the post-disaster recovery process. *International Journal of Disaster Risk Science, 11*, 1–14. https://doi.org/10.1007/s13753-020-00277-8

Montserrat Volcano Observatory. (1997). Definition of boundary limits and volcanic risk map for Soufrière Hills Volcano. https://www.geo.mtu.edu/volcanoes/west.indies/soufriere/govt/miscdocs/rskzone.html

Montserrat Volcano Observatory. (2023). 27th report of the Scientific Advisory Committee on Montserrat volcanic activity. https://www.mvo.ms/wp-content/uploads/2023/08/SAC27_Full_Report_Final.pdf

Movement for Change and Prosperity Party. (2024). 2024 election manifesto. https://montserratfocus.com/wp-content/uploads/2024/10/MCAP-Manifesto-2024.pdf

People's Democratic Movement Party. (2024). 2024 election manifesto. https://montserratfocus.com/wp-content/uploads/2024/10/PDM-Manifesto-2024.pdf

Positive Progression for People. (2024). 2024 election manifesto. https://montserratfocus.com/wp-content/uploads/2024/10/ppp-Manifesto.pdf

Scientific Advisory Committee. (2025). Montserrat volcano observatory. https://www.mvo.ms/

Scolobig, A. (2015). Brief communication: The dark side of risk and crisis communication: legal conflicts and responsibility allocation. *Natural Hazards and Earth System Sciences, 15*(6), 1449–1456. https://doi.org/10.5194/nhess-15-1449-2015

Shotte, G. (2007). Diasporic transnationalism: Relocated montserratians in the UK. *Caribbean Quarterly, 53*(3), 41–69. https://doi.org/10.1080/00086495.2007.11672327

Sword-Daniels, V., Wilson, T. M., Sargeant, S., Rossetto, T., Twigg, J., Johnston, D. M., & Cole, P. D. (2014). Consequences of long-term volcanic activity for essential services in Montserrat: Challenges, adaptations and resilience. *Geological Society, London, Memoirs, 39*, 471–488. https://doi.org/10.1144/M39.26

United Alliance Party. (2024). 2024 election manifesto. https://montserratfocus.com/wp-content/uploads/2024/10/United-Alliance-Manifesto.October-2024_DIGITALpdf_241016_074549.pdf

Wadge, G., & Aspinall, W. P. (2014). A review of volcanic hazard and risk-assessment praxis at the Soufrière Hills Volcano, Montserrat from 1997 to 2011. *Geological Society, London, Memoirs, 39*, 439–456. https://doi.org/10.1144/M39.24

Watson, W. T. (2018). *Feasibility assessment of parametric insurance for volcanic unrest and volcanic eruption: Inception report*. World Bank. Retrieved from https://documents.worldbank.org/en/publication/documents-reports

Wilkinson, E. (2015). Beyond the volcanic crisis: Co-governance of risk in Montserrat. *Journal of Applied Volcanology, 4*(1), 3. https://doi.org/10.1186/s13617-014-0021-7

The manufacturer's authorised representative in the EU is Springer
Nature Customer Service Centre GmbH, Europaplatz 3, 69115 Heidelberg,
Germany. If you have any concerns regarding our products, please
contact ProductSafety@springernature.com

Printed and bound by CPI Group (UK) Ltd, Croydon, CR0 4YY
21/09/2026
02220010-0009